T0401037

ENERGY SCIENCE, ENGINEERING AND TECHNOLOGY

ENERGY RESOURCES

DEVELOPMENT, DISTRIBUTION AND EXPLOITATION

ENERGY SCIENCE, ENGINEERING AND TECHNOLOGY

Additional books in this series can be found on Nova's website under the Series tab.

Additional E-books in this series can be found on Nova's website under the E-books tab.

ENERGY POLICIES, POLITICS AND PRICES

Additional books in this series can be found on Nova's website under the Series tab.

Additional E-books in this series can be found on Nova's website under the E-books tab.

ENERGY SCIENCE, ENGINEERING AND TECHNOLOGY

ENERGY RESOURCES

DEVELOPMENT, DISTRIBUTION AND EXPLOITATION

ENNER HERENIO DE ALCANTARA
EDITOR

Nova Science Publishers, Inc.
New York

For permission to use material from this book please contact us:
Telephone 631-231-7269; Fax 631-231-8175
Web Site: http://www.novapublishers.com

Additional color graphics may be available in the e-book version of this book.

LIBRARY OF CONGRESS CATALOGING-IN-PUBLICATION DATA

Energy resources : development, distribution, and exploitation / editor, Enner Herenio de Alcbntara.
p. cm.
Includes bibliographical references and index.
ISBN 978-1-61324-520-0 (hardcover : alk. paper) 1. Power resources. 2. Power resources--Brazil. I. Alcbntara, Enner Herenio de.
TJ163.2.E4837 2011
333.79--dc23
2011013681

Published by Nova Science Publishers, Inc. † New York

CONTENTS

PREFACE

There is an increase in the projection of future electricity demand across the globe. In the projections from the International Energy Outlook 2010 (IEO2010), the total energy consumption will increase by 49% from 2007 to 2035 (1.4% per year). Based on those reports, it is seen that the use of energy will increase over that time (considering world oil prices will remain relatively high through most of the projection period). There is also an effort being made to search for sustainable and clean energy, like as generated by wind, solar, wave, tide, geothermal and hydroelectricity. This book takes in consideration the following themes: wind power generation, bioenergy, nuclear energy, hydroelectric energy and energy transportation.

Chapter 1 - In order to minimize the human contribution to climate change, scientists have addressed the need for a reduction of present greenhouse gases (GHG) levels by 60 to 80% near 2050. Substantial changes in the energy sector will be required to achieve these levels of reduction. Within this scenario, wind energy emerged as a promising renewable alternative, with proven technology and low environmental impacts.

Chapter 2 - Among the different renewable energy sources (RES) bioenergy is of crucial importance for the current and future energy supply in Central Europe (CE). Not only because it already has the highest share of all RES, but also due to the vast potentials of biomass and the fact that it can be used in all energy sectors: for heat-only, electricity or combined heat and power generation as well as for the production of transport fuels.

Chapter 3 - A nuclear power plant (NPP) considered as a whole presents some by-products that can be exploited to create interesting synergies between the plant itself and potential nearby facilities. These by-products can be ascribed directly or indirectly to the nuclear plant:

- Directly if they derive from the nuclear reactor itself;
- Indirectly if they are due to the presence of the nuclear power station and, consequently, to the features of the location, the need of ancillary installations or the safety measures required by the NPP.

Chapter 4 - The plant cell wall (cell wall herein) is an extra-cellular matrix immediately close to the plasma membrane the components of which are secreted by the plant cell. The cell wall is rigid and attached to the cell wall of the neighbouring cell(s), providing plant cells and tissues with mechanical features that are crucial for life. For instance xylem, the most important vascular tissue for water transport in vascular plants (over 90 % of the water transpired in leaves is transported through xylem), is essentially made of the remaining modified plant cell walls of dead cells. The mechanical properties of xylem cells (tracheids and vessel elements) allow this tissue to stand the strong negative pressures that drive water movement.

Chapter 5 - The energy production to fuel the contemporary global economy has a massive participation of non-renewable sources, accounting for substantial portion of the greenhouse gases released to the atmosphere. According to Friedlingstein et al. and the Global Canopy Programme in 2009 a total of 8.4±0.5 PgC was emitted to the atmosphere by fossil fuel burning, being coal the largest portion of the emissions. This amount accounts to more than 85% of all anthropogenic carbon annually emitted as CO_2 to the atmosphere. With the emissions reduction observed in developed countries, the emissions numbers reflect increase of energy production in developing countries, especially in India and China. The challenge to reduce carbon emissions in the developing world is to achieve the targets without placing the legitimate development goals at risk. A distinct, but internationally recognized, action for greenhouse gases (GHG) mitigation in developing countries is the Nationally Appropriate Mitigation Actions (NAMAs). In Brazil a recent government communication to the United Nation Framework Convention on Climate Change (UNFCCC) states as a national mitigation action the "increase in energy supply by hydroelectric power plants, with estimated reduction of 79 to 99 million tons of $CO_{2\text{-eq.}}$ in 2020".

Chapter 6 - The population of the Republic of Karelia is about 688 000 inhabitants, with over 75% living in the urbanized areas. There are three main towns in Karelia: Petrozavodsk (283 000 inhabitants), Kostomuksha (32 500 inhabitants) and Sortavala (20 200 inhabitants). The population density in the Republic of Karelia is only 4 inhabitants per km^2 (The Republic of Karelia in brief 2010). For comparison, in Finland the population density is 16 inhabitants per km^2.

In: Energy Resources
Editor: Enner Herenio de Alcantara

ISBN: 978-1-61324-520-0

Chapter 1

ASSESSMENT OF THE WIND POWER POTENTIAL OF HYDROELECTRIC RESERVOIRS

Arcilan Assireu[1,*], Felipe Pimenta[2] and Vanessa Souza[1]
[1]Federal University of Itajubá (UNIFEI) / Itajubá, MG, Brazil
[2]Federal University of Rio Grande do Norte (UFRN)/ Natal, RN, Brazil

ABSTRACT

Problems related to the construction of hydroelectric dams have motivated different strategies to mitigate its inherent impacts. One is the multiple uses of dams, which have been commonly employed for recreation, fish farming and flood control. Another possibility not yet investigated, is their potential for harnessing wind power. This alternative is very promising for a couple of reasons.

First, the flooded river valley tends to work as a region of convergence of the air flow, typically yielding higher and steadier winds over water. Second, the wind power transmission is easy to setup as the hydroelectric generation is already connected the power grid. Lastly and more importantly, these renewable forms of energy are complementary. Hydroelectric dams can serve as energy reservoirs, saving water during periods of high wind power generation, and latter produce steady power generation in periods of null winds.

[*]E-mail: arcilan@unifei.edu.br; Tel: +55 35 3629-1521.

This study investigates the wind energy potential of hydroelectric dams of Brazil, one of the largest hydropower producers of the world. Here, a method based on the wind duration curve was employed to determine the persistence of winds for eight tropical reservoirs. A topographic database from three regions (Southeast, Midwest and North) was used to characterize reservoirs relief, i.e. the shape, height and spatial distribution of their roughness elements (mountains and valleys).

Results indicate that Brazilian reservoirs are favorable for the installation of wind farms. Due to their elongated shape, winds tend to blow predominantly aligned with the reservoirs major axis, with persistent (>70% of time under turbine activity) and strong wind speeds (> 5m s^{-1}).

The wind seasonality further suggests they can be highly complementary with hydroelectric power. We found that the stronger (weaker) winds usually occur during the dryer winter (wetter summer) season for any of these dams. This demonstrates that wind farms can significantly contribute to the management of hydropower generation.

Keywords: renewable energy, wind power, hydropower, climate change.

INTRODUCTION

In order to minimize the human contribution to climate change, scientists have addressed the need for a reduction of present greenhouse gases (GHG) levels by 60 to 80% near 2050 (IPCC, 2007). Substantial changes in the energy sector will be required to achieve these levels of reduction. Within this scenario, wind energy emerged as a promising renewable alternative, with proven technology and low environmental impacts (Archer and Jacobson, 2005).

Because of that, wind is presently the fastest growing energy source in the world, with more than 158 GW installed capacity (GWEC, 2009). In Brazil, wind energy accounts for only 0.72% while hydropower is responsible to approximately 70% of the total energy generation (ANEEL, 2010).

Although hydroelectric power constitutes a valuable asset for Brazil, the construction of dams has resulted in several ecological and societal problems (Fearnside, 2004). The government has established a few strategies to mitigate these impacts. The multiple use of reservoirs, for example, have encouraged their employment for recreation, fish farming and flood control. Another possibility not yet investigated, is the potential of hydroelectric dams for harnessing wind power. While the Brazilian wind power potential over land (Amarante et al., 2001;

Feitosa et al., 2003) and for some offshore regions (Pimenta et al., 2008) has been previously estimated, the assessment of the wind potential for lakes and artificial damns was not studied.

Different works have addressed the dynamics of the airflow over lakes and other large bodies of water. They found that differences on the surface drag between the land and water bodies, and the effect of complex topography adjacent to reservoirs will lead to large spatial variations of the winds (Whiteman and Doran, 1993; Guardans and Palomino, 1995; Offer and Goossens, 1994; Bullard et al., 2000). In similar way, the differential warming that occurs between land and water will lead to convection and the generation of a "breeze" of marked diurnal variability (Huthnance, 2002).

The topography adjacent to reservoirs causes the convergence and alignment of the airflow along the direction of the valley. The result is an increase on the winds mean speed over water and a reduction of the wind directional variability (Mason, 1986, Hunt et al., 1988a; Hunt et al., 1988b; Finnigan et al., 1990; Bullard et al., 2000). Both effects are important for windpower generation and, in specific, the reduced directional variability facilitate the micro sitting of wind turbines. The valley circulation, however, is naturally susceptible to the effects of large-scale atmospheric systems (Weigel and Rotach, 2004; Bitencourt et al., 2009).

The alternative of harnessing wind power over Brazilian hydro-electrical reservoirs is promising for a couple of good reasons. First, the flooded river valley can work as a region of convergence yielding stronger and steadier winds. Second, the energy transmission is reasonably simple to setup, as the hydroelectric generation is already connected to the power grid. Lastly and more important, both renewable forms of energy can be complementary. While hydroelectric reservoirs provide favorable conditions for wind power generation, the energy from wind turbines can help on the management of dams' water levels through the reduction of hydropower generators' activity. That is, the water saved in periods of high winds can be used for hydroelectric generation during periods of null winds or even during dryer climate conditions.

This study explore the characteristics of winds from eight hydroelectric reservoirs of Brazil in order to assess their potential for electricity generation. Meteorological buoys are combined with a topographic dataset to set the timing, strength, persistence of winds and the characteristics of their surrounding relief and roughness elements. Section 2 describes the Data and Methods used for the current work. Section 3 report and discuss the Results. Finally, we offer a Summary and Conclusions on Section 4.

Data Set and Methods

Study Area

Eight hydroeletric reservoirs (Furnas, Estreito, Manso, Funil, Itumbiara, Corumbá, Serra da Mesa and Tucurui) are included in this study (Figure 1). Considering the first seven reservoirs mentioned above, prior to flooding the original vegetation was the Cerrado Biome (Savannah), which is situated on poor soils overlying pre-Cambrian rocks. The regional climate is Tropical Humid (Niemer, 1989), with a rainy season during summer and a dry period in the austral winter (June–August), when the average temperature for the coldest month is above 18°C. All reservoirs are used for the generation of electricity, and their average annual output range from 180 MW to 8.370 MW (Table 1).

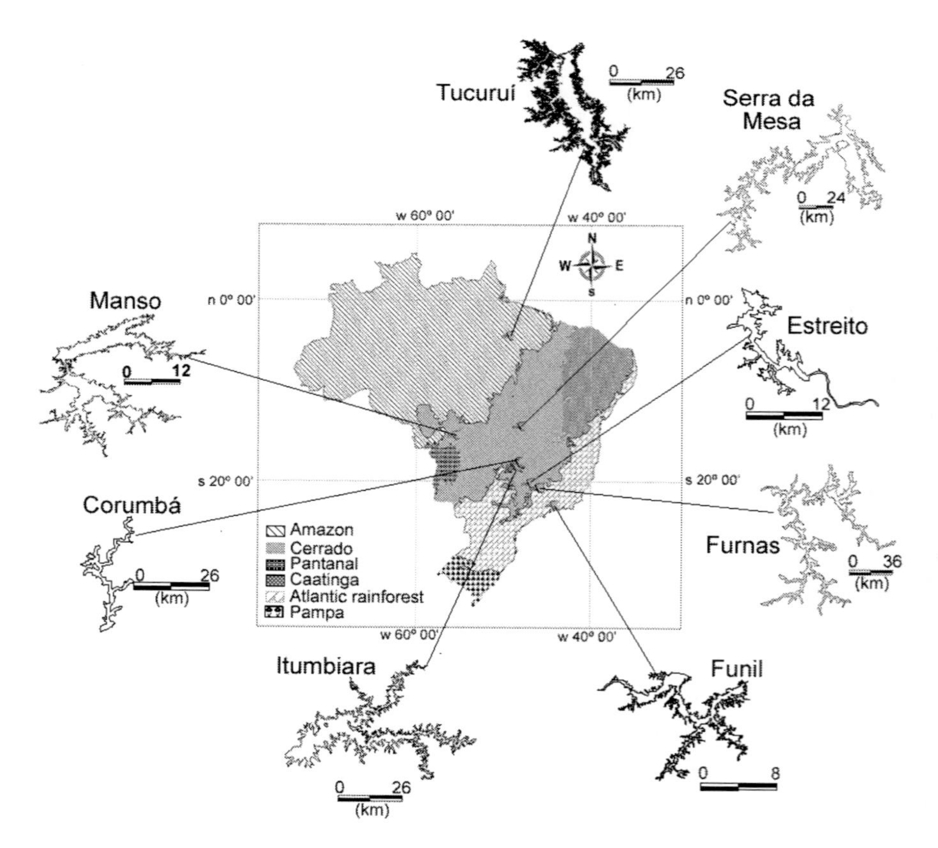

Figure 1. Location of the studied aquatic systems. Biome types are indicated by the legend.

Table 1. General features of the eight reservoirs

Reservoir	Maximum storage (m)	Minimum level of operation (m)	Flooded area (km^2)	Total volume (m^3)	Useful volume (m^3)	Power (MW)
Corumbá	595	570	65	1.5×10^6	1.0×10^6	184
Estreito	622	618	47	1.4×10^9	0.2×10^9	1050
Funil	466	444	40	8.9×10^9	6.2×10^6	180
Furnas	768	750	1440	23×10^9	17.2×10^9	1216
Itumbiara	520	495	778	17×10^9	12.4×10^6	2080
Manso	287	245	427	7.3×10^9	2.9×10^6	210
S. Mesa	460	417	1784	54×10^9	43.2×10^6	1275
Tucuruí	-	-	2785	50×10^6	35.3×10^6	8370

The Tucurui reservoir is located in Amazonia (Figure 1). The regional climate on Amazonia represents a combination of several factors, where the most important is the availability of solar energy through energy balance. Due to high amounts of solar radiation, the air temperature shows little variation throughout the year. The seasonal temperature range is about 1-2°C, and the average values are between 24 and 26°C. This region has an average rainfall of about 2300 mm $year^{-1}$, although the annual total may reach 3500 mm $year^{-1}$ in some regions near the border between Brazil, Colombia and Venezuela (Niemer, 1989).

Meteorological Dataset

The wind data set was collected by an anchored buoy named Environmental Monitoring Integrated System (SIMA), an equipment projected for data acquisition and real time monitoring of hydrological systems (Stevenson et al. 1993). The SIMA is composed by an anchored buoy, where meteorological (air temperature, wind speed and direction, humidity, pressure, incoming and reflected radiation) and water sensors (a thermistor chain) are installed. The system also counts with a data storage unit, solar panels, batteries and a transmission antenna (Figure 2). The data is collected at 3 hours time interval and is transmitted by satellite for a reception station. Winds, air temperature and radiation are measured at 3 m height (data available, under demand, at www.dpi.inpe.br/sima).

THE POWER LAW PROFILE

Meteorological wind data are taken near the surface, or at meteorological buoy height (3 m). In wind energy studies, we are usually interested in wind at the height of the hub of a wind turbine (70–100 m). There are several methods to extrapolate winds that are based on models of differing complexity of the Planetary Boundary Layer (PBL). Some of these methods are reviewed in the Appendix of this Chapter. For the practical purposes of this study, however, we make the use of the Power Law Profile (Justus and Mikhail, 1976):

$$V(z) = V_a \left(\frac{z}{z_a} \right)^n \quad (1)$$

where V(z) is the velocity at a given height z, z_a is the height of our measured wind speed (Va), and n is an exponent whose value depend on the measured wind speed and the reference height.

Figure 2.Photo of the Environmental Monitoring Integrated System (SIMA).

$$n = \frac{0.37 - 0{,}088\ \ln V_a}{1 - 0.088\ \ln z_a / z} \tag{2}$$

For other formulations, this parameter might also depend on the atmosphere stability or the surface roughness (see Appendix).

PROBABILITY DISTRIBUTION FUNCTION

Several studies have indicated that the Weibull Probability Density Function (PDF) gives the best fit to wind data (Van der Auwera et al., 1980; Rehman et al., 1994; Garcia et al., 1998; Nfaqui et al, 1998; Silva et al., 2002, Archer and Jacobson, 2003). Its PDF is given by:

$$p(V) = \left(\frac{k}{c}\right)\left(\frac{V}{c}\right)^{k-1} \exp\left[-\left(\frac{V}{c}\right)^{k}\right], \tag{3}$$

and its determination requires knowledge of two parameters: k, a dimensionless shape factor and c, a scale factor (velocity units). Both are functions of the mean wind speed $\overline{V}$ and its standard deviation σ. Justus et al. (1978) tested several methods for estimating the shape and the scale factors. Among them, the most suitable for the typical wind intervals observed in the reservoirs previously mentioned are:

$$k = \left(\frac{\sigma}{\overline{V}}\right)^{-1.086} \tag{4}$$

$$c = \frac{\overline{V}}{\Gamma(1 + 1/k)}, \tag{5}$$

where Γ is the gamma function.

Justus and Mikhail (1975) also suggest the following empirical relationships which allows to adjust the Weibull distribution to different heights z, based on the factors c_a e k_a, derived from Equation (4-5) to an anemometer height z_a:

$$c(z) = c_a \left(\frac{z}{z_a} \right)^n \qquad (6)$$

$$k(z) = k_a \frac{\left[1 - 0{,}088 \ln\left(\frac{z_a}{10} \right) \right]}{\left[1 - 0{,}088 \ln\left(\frac{z}{10} \right) \right]} \qquad (7)$$

where z e z_a are in meters and the exponent of the power law n is given by:

$$n = \frac{0{,}37 - 0{,}088 \ln c_a}{1 - 0{,}088 \ln z_a / 10} \qquad (8)$$

Fitness of the Weibull PDF

The fitness of the Weibull PDF distribution can be tested based on the Kolmogorov-Smirnov (Massey,1951) adherence test. The test measures the degree of agreement between the distributions of a set of sampling datobservations $p_{obs}(V)$ and a given theoretical distribution $p(V)$:

$$KS = max\left| p_{obs}(V) - p(V) \right| \qquad (9)$$

The KS value is then compared to a statistical table for the Kolmogorov-Smirnov test, which is a function of the sample size and a defined significance level. If KS is larger or equal than this tabulated value, one can affirm that there is a good agreement between the theoretical and observed distributions.

Wind Speed Persistence

Wind speed persistence is a useful measure for wind energy studies, which is typically assessed through the calculation of speed duration curves (SDC) (Koçak, 2002). Good sites for wind exploration tend to demonstrate frequent conditions when winds have enough speed for energy production, which are winds above the turbine cut-in speed. The SDC derives from the cumulative sum of occurrences of wind speeds equal or higher than a specific magnitude. In other words, it is the graphical presentation of wind speed versus the percentage of time for which each speed equals or exceeds this particular value. A SDC representation, with allowance for wind direction is given by the following equation:

$$S(i,j) = \frac{1}{N}\sum_{i=1}^{N} H[V(t,\emptyset) - i\Delta V(i,j)] \tag{10}$$

where i represents the classes of wind speed and j=1,2,....D is the directional partition of bins. Note that D = 360°/Δϕ, where Δϕ is a prescribed angle partition. $S(i,j)$ represent the percentage of time that winds are equal or above a particular value, $V(t,\phi)$ is the hourly averaged wind speed (m s^{-1}) at time t and from direction ϕ, and $\Delta V(i,j)$ is a predetermined wind speed increment. The Heaviside step function is represented by H(x), being 1 for x ≥ 0 and 0 for x <0. On this equation the maximum number of classes is given by $N = (V_{max} - V_{min})/\Delta V$. If $S(i,j)$ values are plotted versus $i\Delta V(i,j)$ values the SDC for the direction j is obtained.

RESULTS AND DISCUSSION

Winds Probability

In most situations wind speeds are fairly modeled by the Weibull PDF. This statistical distribution enables us to estimate how often winds of a specific speed will be found based on their average $\overline{V}$ and standard deviation σ. Here this probability distribution function is computed for all reservoirs. For illustration purposes, we describe in more detail the results for the Itumbiara reservoir. Figure 3a. compares the observed (vertical bars) and theoretical (continuous line) probability distribution functions for Itumbiara. The mean wind speed (at the height of 100 m) is 7.90 m s^{-1}. A vertical dashed line in the same graph marks the median wind speed at 7.60 m s^{-1}. The median separates the data in two groups, one where 50% of time winds speeds are lower than the median and, the other, where winds are higher than the median for 50% of time. The asymmetry of the curve, however, evidences that low and moderate winds can be more common in contrast to strong events, which are relatively rare. What controls the asymmetry of these curves is the Weibull shape parameter k. High values of k are associated with high medians, while locations of wide distributions (lots of low and high winds) would have $k<2$, being $k=2$ the limit case of a Rayleigh distribution. As seen by Table 2, most of the hydroelectric reservoirs presented $k \leq 2$, while some locations presented fairly consistent wind speeds with $k \approx 3$, as observed for Itumbiara (Table 2).

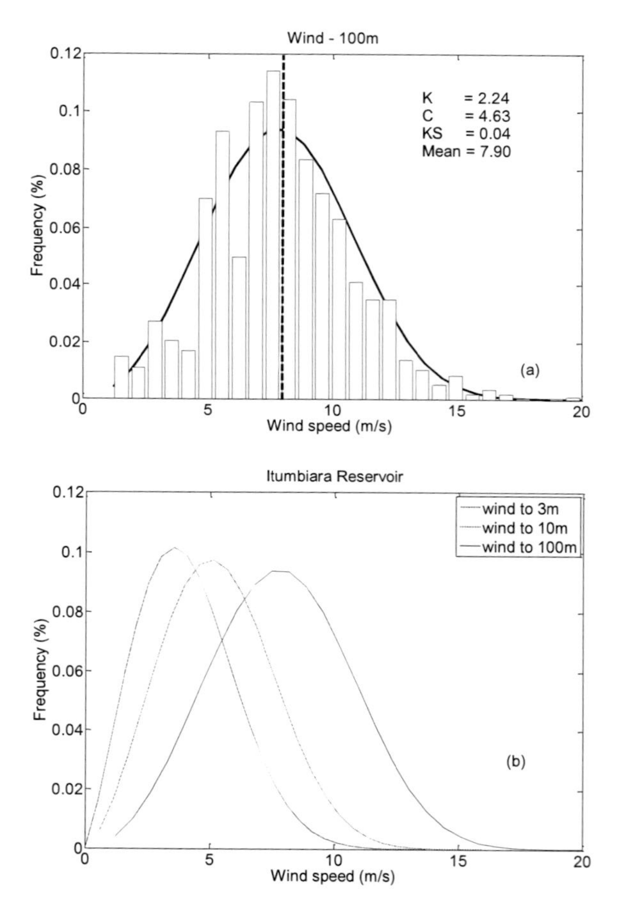

Figure 3. (a) The Weibull probability distribution function for Itumbiara reservoir. Bars represent observations and the continuous line the theoretical distribution. The median is indicated by a vertical dashed line (b) Example of Weibull PDF computed from Equations (6-10) for the heights of z=3, 10 and 100m.

In terms of mean speed, the other reservoirs range between 4.2 and 5.8 m s^{-1} with class rankings varying between 1 and 4 (Table 2). Figure 3.b. further illustrates the changes of the Weibull PDF for increasing heights at Itumbiara. The blue curve in this graph represents the distribution for the anemometer height (z = 3 m), while the red and black lines respectively show the distributions for 10 and 100 m. It can be noticed that these curves are displaced to the right (toward higher wind speeds) and that their distribution becomes wider for increasing heights.

The goodness of fit between theoretical and observed probability distributions is typically represented by low values of the Kolmogorov-Smirnov test (Sansigolo, 2005). As seen on Table 2 KS values computed for our reservoirs are generally lower than 0.1, the best statistical adjustment (KS=0.04) occurring for Estreito, Itumbiara and Manso Reservoir. All stations distributions passed the Kolmogorov-Smirnov test.

Periods of Turbine Activity

The consistency of electrical output can be assessed by examining the percentage of time under turbine activity. Present modern turbines have a minimum speed around 3 m s^{-1}, called the “cut-in” speed, below which they do not produce power. They also have a maximum or “cut-out” speed above which they shut down for self-protection, also not producing power (typically around 25 to 30 m s^{-1}). The period of activity is characterized therefore as the percentage of time with winds speeds between the cut-in and cut-out speeds.

The period of activity was computed for each reservoir from all wind speed data available, regardless of the direction winds were blowing. Results revealed that wind turbines could be generating power at least 70% of time in any reservoir (Table 2). For Itumbiara in specific, the turbine activity was very high: 93% of time this site would be producing wind power, but not at a constant output. Thus these wind sites are not properly considered when called intermittent. A better description would be that they produce varying output, in contrast to a hydro-driven power plant, whose output is controlled by the operator. The wind speed duration curve for Itumbiara is shown on Figure 4. This distribution is derived from Equation 10, considering all wind speeds (and all directions) available. The horizontal dashed lines in this graph further represent typical turbine cut-in and rated speeds. The crossing of these lines with the duration curve therefore indicates the percent of activity and rated power respectively. In this example, it is shown that nearly 5% of time Itumbiara would achieve maximum power production. The distribution further shows that winds never reach cut-out speeds for the observed period.

Wind Power Density

In order to estimate the basic size of the resource, another quantity of interest is the wind power density $P = (\rho_{air} V^3)/2$, where ρ_{air} is the air density (typically 1.165 kg m^{-3} for tropical regions) and V is the wind speed. A typical time-series of wind power density is shown on Figure 4.b. for the Itumbiara reservoir. The graph shows many ups and downs related to winds' variability. For most turbines, these peaks would be leveled out because turbines achieve a constant output after winds reach their rated speed of operation. The mean power density observed for Itumbiara was P=405 W m^{-2}, which is comparable to the magnitude observed for Brazilian offshore regions (446 to 775 W m^{-2}) (Pimenta et al., 2008[1]).

Table 2. Wind properties at each reservoir. Record length (months); $\overline{V}$ and σ are respectively the mean and standard deviations of wind speed at 100 m height (m s^{-1}); $\overline{P}$ is the mean power density (W m^{-2}), *k* and *c* are the shape and scale parameters, while KS is the Kolmogorov-Smirnov test. Class refer to wind classes similar to Archer and Jacobson (2005). %Active refers to the period of time that winds are higher than 3 m s^{-1} (i.e. a typical turbine cut-in speed)

Reservoir	months	$\overline{V}$	σ	$\overline{P}$	k	c	KS	Class[2]	% Active
Corumbá	12	4.5	2.3	98	1.9	2.12	0.06	1	75
Estreito	12	5.8	2.4	176	2.5	3.12	0.04	2	90
Funil	12	4.2	2.3	85	1.8	1.85	0.09	1	70
Furnas	12	5.8	2.9	201	1.9	3.10	0.06	2	80
Itumbiara	20	7.9	2.8	405	2.9	4.60	0.04	4	93
Manso	18	5.7	2.7	186	2.1	2.90	0.04	2	90
Serra da Mesa	52	4.5	2.2	95	1.9	2.20	0.07	1	75
Tucuruí	47	4.7	2.8	133	1.7	2.03	0.13	1	70

[1]Pimenta et al. (2008) considered in their analysis, however, the hub height of z=80 m.

[2]Here we consider slightly different class intervals from Archer and Jacobson (2005). Class 1: V<5.7ms-1; Class 2: $5.7 \leq V < 6.9$ ms-1; Class 3: $6.9 \leq V < 7.5$ ms-1; Class 4: $7.5 \leq V < 8.1$ ms-1; Class 5: $8.1 \leq V < 8.6$ ms-1; Class 6: $8.6 \leq V < 9.4$ ms-1; Class 7: V > 9.4 ms-1.

The Modification of Airflow by Local Topography

In order to evaluate the influence of topography, we investigated the reservoirs' relief in combination to the directional distribution of winds. We describe the topographic relief from the Shuttle Radar Topographic Mapper (SRTM) database (Farr et al., 2007). An example of a SRTM map is shown on Figure 5.a,b. for Itumbiara. This reservoir was constructed by damming rivers previously flowing between mountains, so that the flooding of the river valley resulted on an elongated shape for the water body. For this reservoir, in particular, the mountains tend to channel the winds, providing the convergence of low level jets associated with easterly winds.

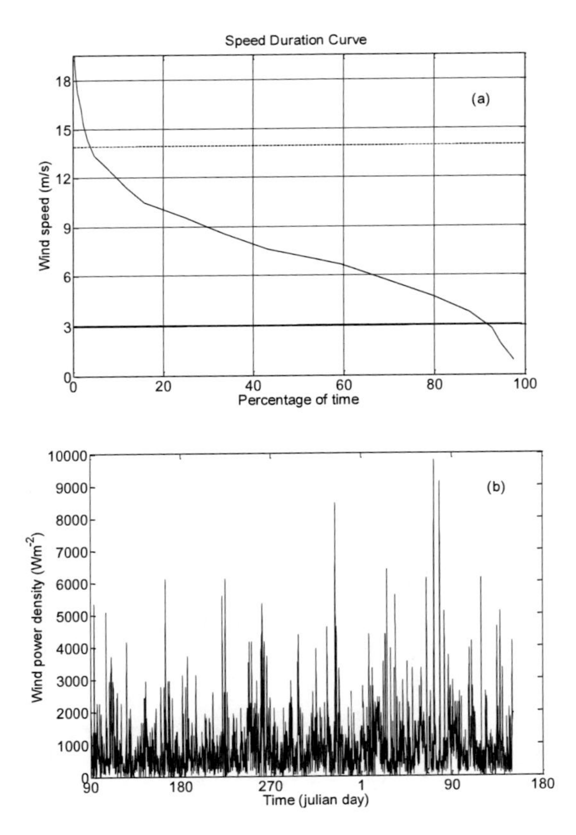

Figure 4. (a) Wind speed duration curve for Itumbiara. The horizontal lines at 3 and 14 m s^{-1} are respectively the cut-in and rated speeds for typical modern turbines. (b) Correspondent time series of wind power density.

Accordingly, the wind rose distribution for Itumbiara, shown on Figure 6.a, illustrate that easterly winds are the predominant direction for this reservoir. This is in agreement with the reservoirs main fetch direction. Winds from the eastern quadrant typically occur > 35% of time for Itumbiara (Figure 6). Since large-scale winds are approximately parallel to the mountains, the local topography has a funneling effect, modifying the wind direction and the wind speed (i.e. speed-up phenomenon) (Bullard et al., 2000). The process is similar to a mechanism identified by Whiteman and Doran (1993) for the Tennessee Valley, USA. Large-scale winds associated with synoptic systems pass over the valley, converging between the mountains, causing a downstream acceleration of the flow and strong winds over the reservoirs' surface with similar or different direction than the synoptic system.

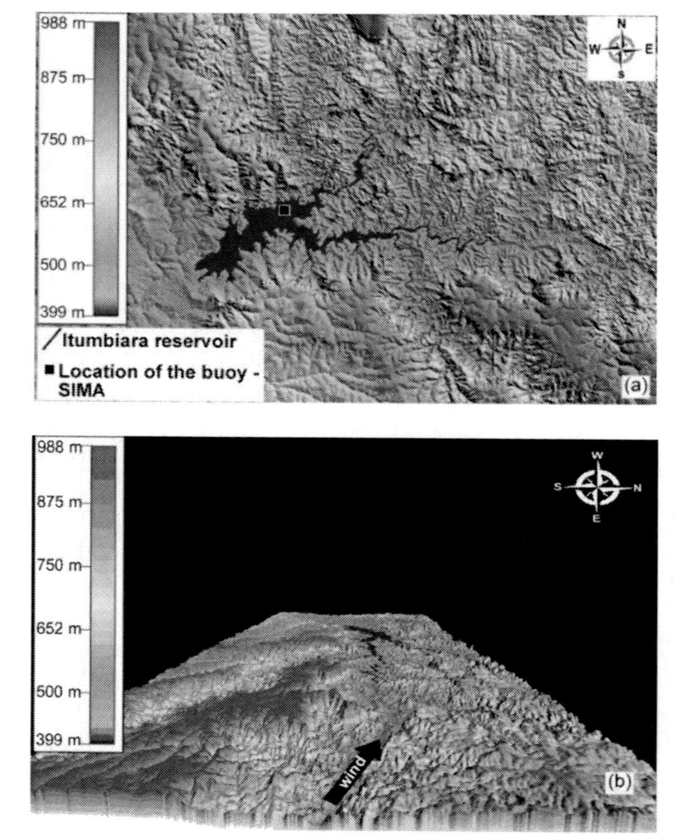

Figure 5. (a) Topography and (b) elevation map of the Itumbiara Reservoir. The arrow on (b) indicates the prevailing wind direction.

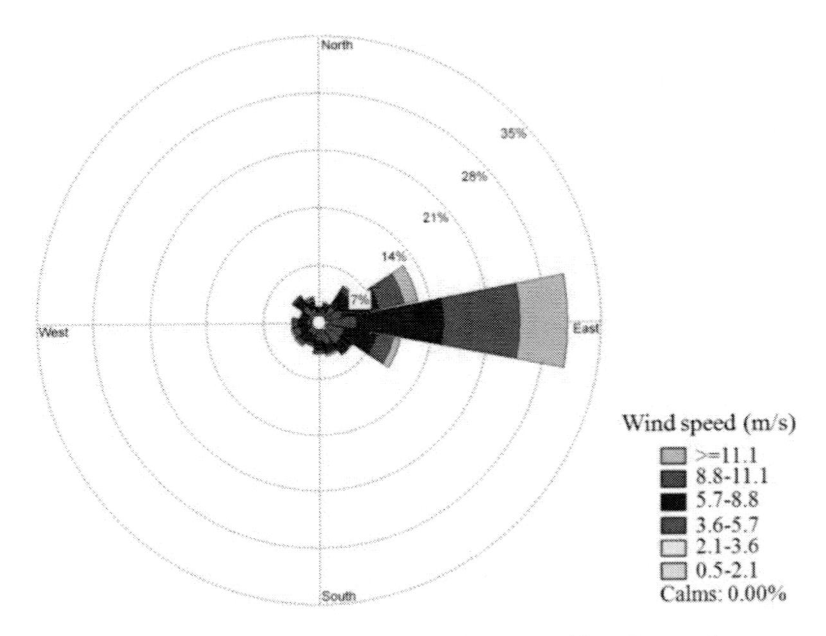

Figure 6. Frequencies of wind directions and speeds for Itumbiara Reservoir.

The same analysis was extended to the other reservoirs and can be summarized as the following. Serra da Mesa, Corumba, and Tucurui (Figure 7 a,c,d) are flooded valleys of moderate topographic elevations. The others (Manso, Tucuruí, Furnas and Estreito) (Figure 7 b,d,e,f) were constructed between high mountains chains and a generally larger fetch, which results on more efficient funneling of large-scale winds. The higher winds' persistence was found, in accordance, for Manso, Itumbiara, Furnas and Estreito (Table 2).

Morphology of the Reservoirs' Valleys

In order to characterize the spatial variations of the surrounding terrain, we investigated the morphology of our reservoirs' valleys by sampling different sections from our topographic database. The Figure 8 illustrates how these sections were selected.

The method consists of first drawing an axis parallel to the prevailing wind direction (as measured by the anemometer), which passes through the SIMA buoy position. Then, we draw two radials at angles $\pm$ 45° from the main axis. (Figure 8a).

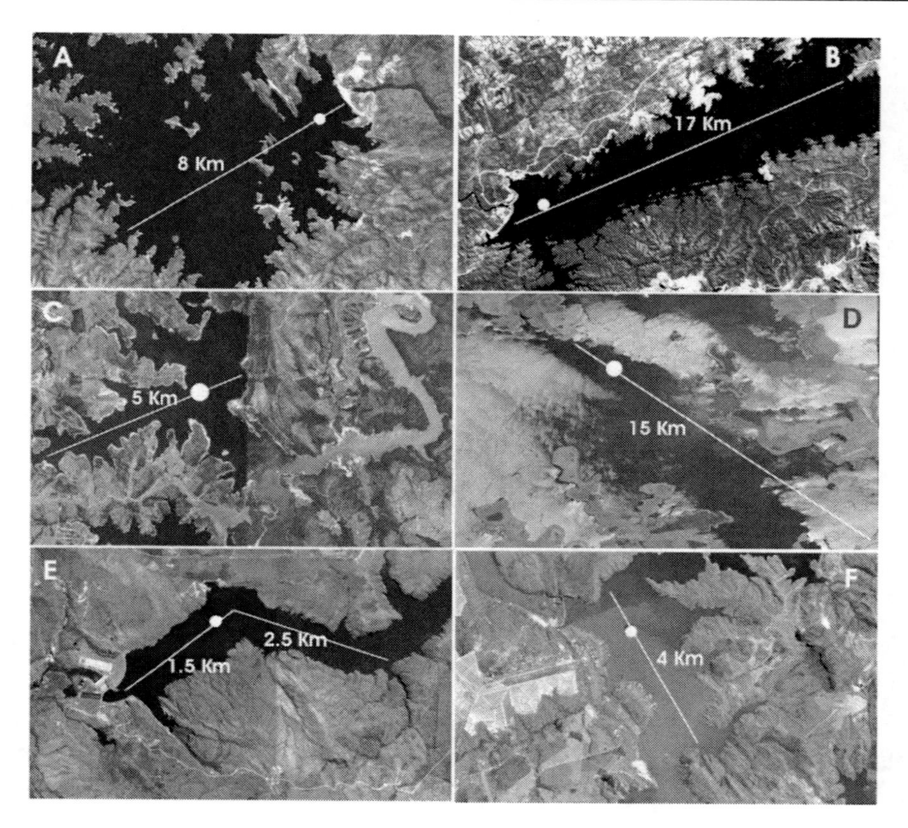

Figure 7. Fetch associated with each reservoir. Filled dots indicate the location of SIMA buoys. (A) Serra da Mesa, (B) Manso, (C) Corumba, (D) Tucurui, (E) Furnas and (F) Estreito.

Next, from the buoy position, we start drawing concentric circles, first at a distance 2L, and then subsequently adding an increment L to the circle's radius (Figure 8a,b).

The topographic sections are then taken perpendicular to the wind axis direction, the first S1 being defined between the points of intersection between the sections and circle of radius 2L (Figure 8.a).

The following sections, S2 (Figure 8.b), S3 (Figure 8.c) and S4 (Figure 8.d), define the intersections to other concentric circles. The results shown on Figure 8. illustrate that two valleys of nearly 300 m depth converge at Itumbiara, what explains why the funneling effect of prevailing easterly winds is so prominent at this location. The same type of analysis illustrates the topography influence over the large-scale atmospheric circulation of Estreito (Figures 9a and 10a-d).

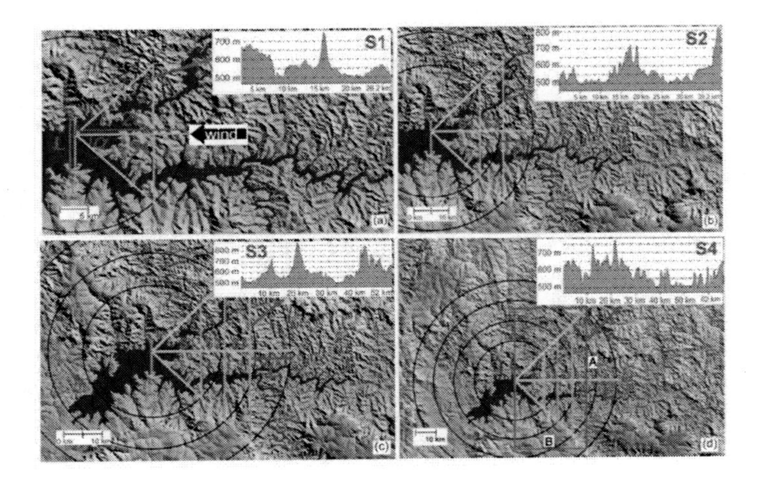

Figure 8. Definition of the topographic sections of the river valley. The sequence from (a) to (d) illustrate the sampling of sections S1 to S4 located upwind of the meteorological buoy (see text for details).

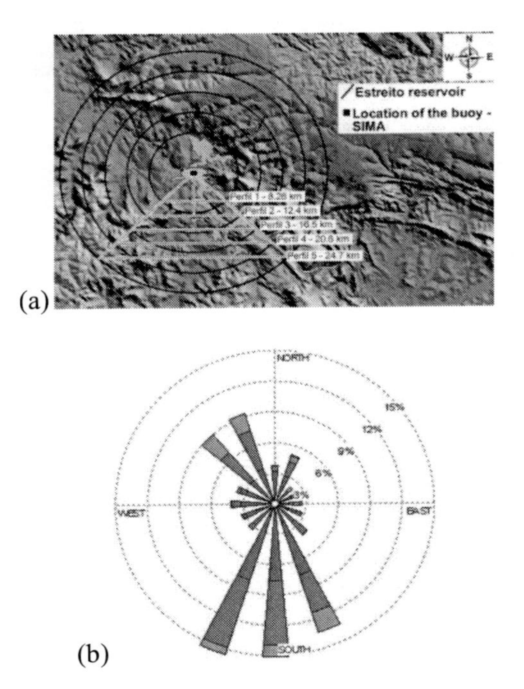

Figure 9. (a) Surrounding topography of the Estreito Reservoir. (b) Directional distribution of winds.

Although easterly winds are the predominant direction from large-scale systems, the observed winds inside Estreito valley are from the south quadrant. (Figure 9.b). An analysis of its topography profiles (Figure 10. a,b,c,d) makes clear the influence of the relief. In this case, the western mountain chain channels the airflow from large-scale systems, guiding it through the river valley main axis, further causing more convergence and stronger winds over the reservoir. Closer to the buoy location, the valley narrows and the hill in the reservoirs' sides provide a tunnel effect, resulting in near parallel winds to the reservoirs axis (see Figure 9.b. and 10.a). This kind of topographic effects can be even observed for reservoirs of Amazonia, which have a notably smoother topography (Figure 11.a). For this reservoir, the large-scale atmospheric circulation is dominated by north and northeast winds, what coincides with the direction of the mountain channel (Figure 11). Thus, the influence of topographic effects can contribute to the exploration of wind energy of Amazonian reservoirs.

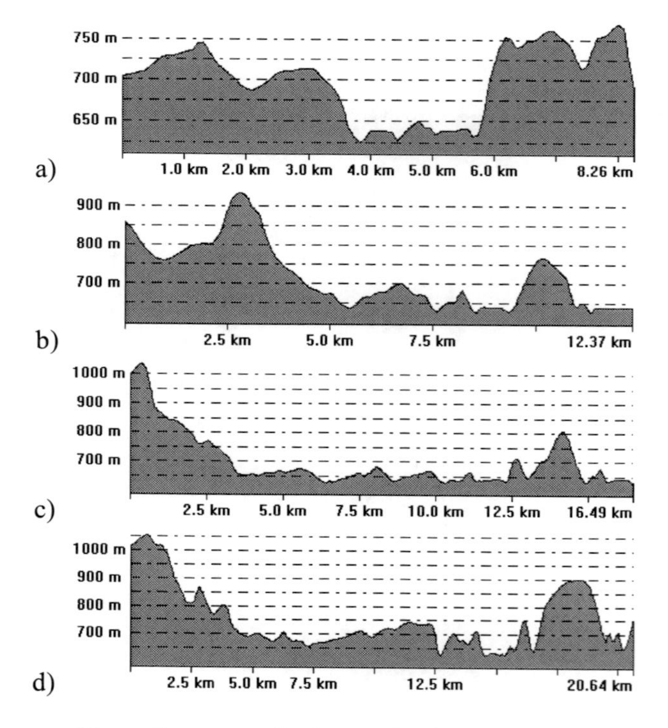

Figure 10. Topographic profiles perpendicular to wind direction for the Estreito reservoir. Sections (a) to (d) are drawn from the meteorological buoy position to the south (see Figure 9. for positions).

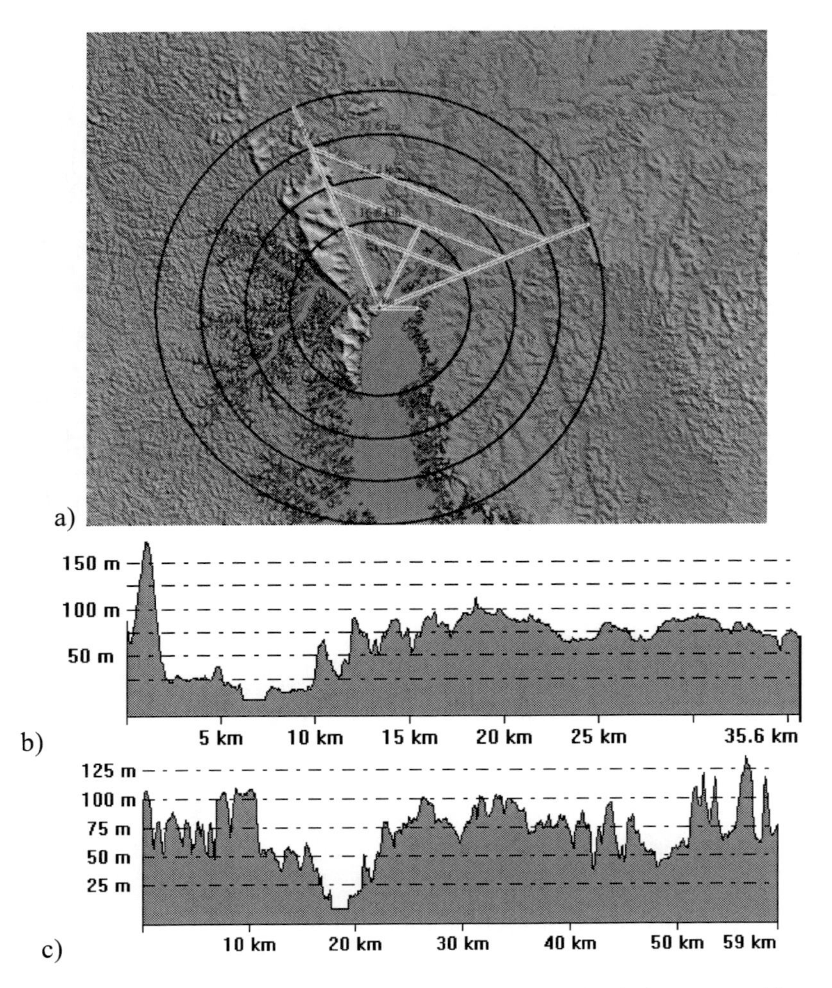

Figure 11. (a) Surrounding topography of the Tucurui Reservoir. (b,c) Topographic profiles perpendicular to wind direction.

Managing Dry Periods with Wind-Hydro Integration

Winds can be highly variable in time and thus hard to integrate to the grid. This is especially significant as wind power becomes a higher proportion of all generation, and it turns more difficult for electric system operators to effectively integrate the fluctuating power output. An efficient way to use (and aggregate value) to wind energy, however, is to associate its use with large storage units, such as hydroelectric reservoirs.

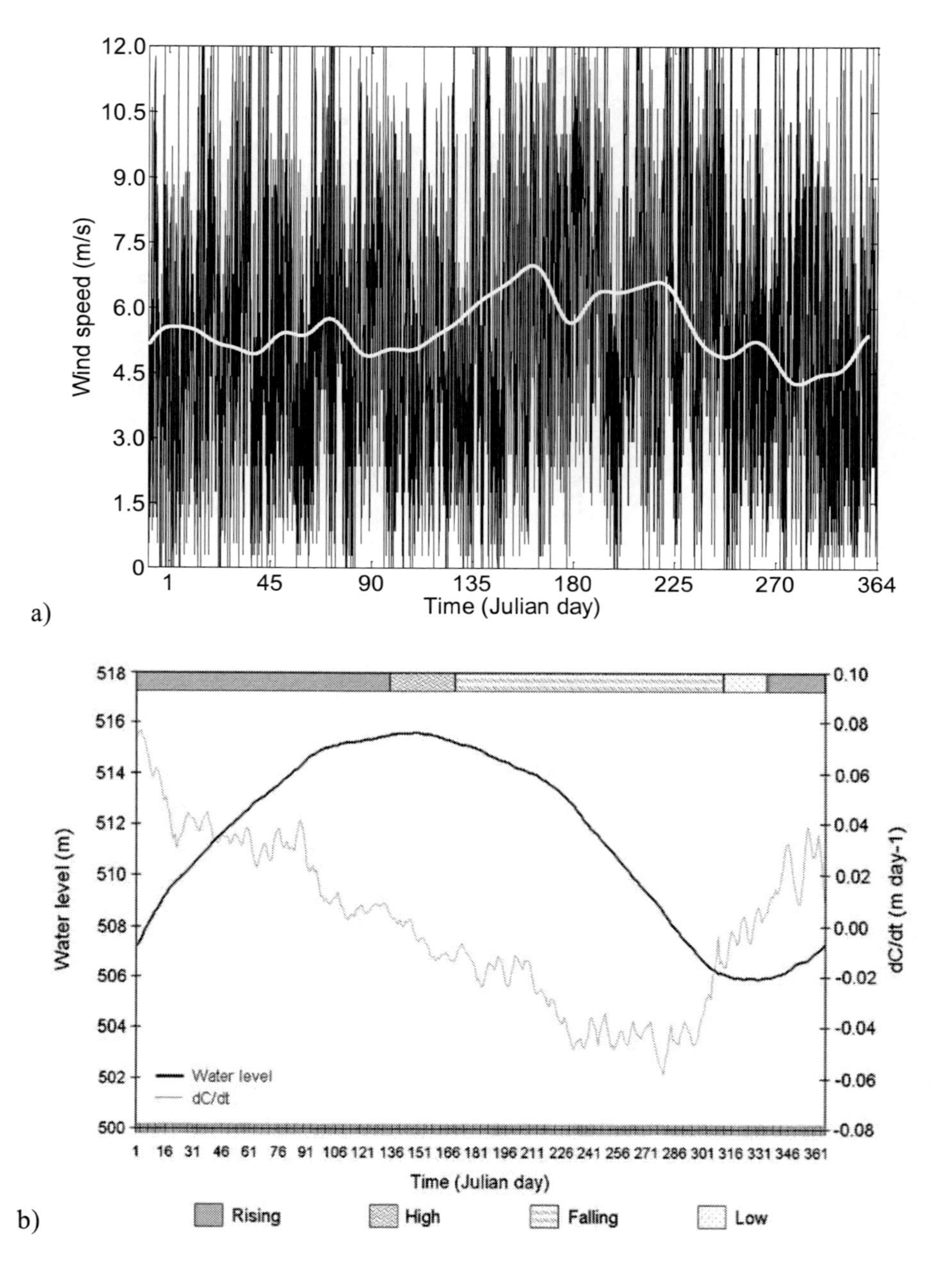

Figure 12. (a) Itumbiara time series of wind speed (thin black line) and its seasonal tendency (white line). (b) Daily averages of water level fluctuations (black line) and its time derivative (gray line) (after Alcântara et al. 2010).

This allows one to estimate more securely how much energy will be generated in a season or a year without worrying about higher frequency (i.e. daily or hourly) fluctuations of a wind farm. The integration should be possible through the management of hydropower plants outflow. When wind turbines are at the peak production hydropower gates can be closed, thus saving water while maintaining constant power output of the combined wind-hydropower system.

This integration can be especially useful for periods of prolonged drought; in many tropical reservoirs winds will naturally tend to be more energetic during the dry season. A practical example for the Itumbiara reservoir is shown on Figure 12. The left panel on this figure shows the wind speed fluctuations (thin black line) and its respective seasonal tendency (white line). One can notice that the period of higher wind speeds occur between the April and October (Julian days 100 and 240). On the right panel, a continuous black line shows Itumbiara reservoir's water level, which illustrate how that the falling cycle of water level coincides with the period of higher winds. The same type of analysis was performed for the other reservoirs and a similar complementarity of winds to the hydrological cycle was found (results not shown).

CONCLUSION

Meteorological data from eight Brazilian hydroelectric reservoirs were analyzed in combination to a topographic information database to address three main objectives: (i) Assess the wind power potential of hydroelectric dams; (ii) Understand how topographic features influence the airflow that blow over the reservoirs' surfaces, and (iii) Explore the seasonal complementarity of wind and hydro power generation.

We found that for any reservoir wind turbines would be providing power (but not at a constant output) for at least 70% of time. Some stations like Itumbiara, remarkably achieved 93% of turbine activity. Wind power densities varied between 98 and 405 W m^{-2} among reservoirs, where the four largest were Itumbiara (405 W m^{-2}), Furnas (201 W m^{-2}), Manso (186 W m^{-2}) and Estreito (176 W m^{-2}). The power density found for Ibumbiara is of comparable magnitude to some offshore stations of south Brazil (Pimenta et al., 2008).

Results indicate that the mountainous relief surrounding these reservoirs can cause a marked deflection of the large-scale atmospheric flow, which will tend to flow more parallel to the reservoirs main axis. This contributes to the steadiness of wind power generation. On the other hand, the funneling effect of certain river valleys combined with the low drag of reservoirs surface cause the convergence of winds, which explain the high mean wind speeds found for Itumbiara (~8 m s^{-1}) and Furnas, Estreito and Manso (~6 m s^{-1}).

We found that hydro and wind power systems are highly complementary throughout the seasons. During dry seasons, tropical reservoirs tend to be quite windy, so that local wind power generation allows the reduction of hydropower's activity and outflow. In similar way, the surplus of energy provided by winds in other seasons can be used in the long-term management of the reservoirs, helping to keep high water levels prior to extended periods of droughts.

The impacts of global climate change scenarios on the vulnerability of the Brazilian energy system in general (Lucena et al., 2010a), and in its wind power system in particular (Lucena et al., 2010b) has been studied, but this effect specifically on the hydroelectric regions has not yet analyzed. The total power generation of this resource and the most suitable regions to wind farm installation inside these reservoirs are under investigation and will be reported in further communications.

ACKNOWLEDGMENTS

We acknowledge the Furnas Centrais Elétricas S. A. for making wind speed data available.

APPENDIX A

The Behavior of the Planetary Boundary Layer for Tropical Reservoirs

The airflow that turns wind turbines is located inside the lowest layer of the atmosphere, which is in close contact with the surface of continents and water bodies, i.e. the Planetary Boundary Layer (PBL). The PBL typical thickness is around 1 km height. Wind turbines, on the other hand, are positioned around 80 to 100 m height, or in the lower PBL. It is in this specific region where the largest variation of winds as a function of height occurs, and where a good knowledge of the physical characteristics of the PBL is necessary.

Historically, the studies on turbulent transport were developed under conditions in which the humidity and the thermal structure of the PBL was not considered important in the process. For example, part of the constants and parameters used in studies involving turbulence were obtained from wind tunnels. But there are at least two differences between the PBL structures over land as compared with water. First, the surface roughness z_0 is typically, at least, an order of magnitude lower over water as compared with land (Hess and Garrat, 2002). This results in higher wind speeds over water. Second, the intense water-air temperature differences in tropical reservoirs result in variable conditions of stability.

The Power Law (Eq. 1) can be adjusted for the inclusion of these processes. Equation (2) can be rearranged as in Panofsky and Dutton (1983) to give

$$n = \left(\frac{z}{V}\right)\left(\frac{\partial V}{\partial z}\right) \quad \text{(A1)}$$

An estimate for the dimensionless wind shear as a function □ dependent of z/L results in (Zoumakis, 1993),

$$\frac{\partial V}{\partial z} = ((V_*/kz)\phi(z/L)) \quad \text{(A2)}$$

The integration of Equation (A2) results in the wind profile for the conditions of stable atmosphere,

$$V(z) = (V_* / \kappa)\{\ln(z / z_0) \quad + \quad \beta(z - z_0)/ L\} \quad \text{(A3)}$$

Where V_* is the friction velocity, κ = 0.40 is the Von Karman constant, L is the Monin-Obukhov parameter, z_0 is the scale of roughness (i.e. a characteristic of the surface), and β = 4.7. A recommended average value for surface roughness is z_0 = 0.2 mm, for conditions of calm surface water (Manwell et al., 2002). A time dependent formulation for z_0 is described by Garvine and Kempton (2008). For the atmospheric boundary layer, considered up to 150m, Hsu (2003) proposes,

$$V(z) = (V_* / k)\{\ln(z/z_0) \quad - \quad \psi_m(z/L)\} \quad \text{(A4)}$$

That can be written as,

$$V(z) = (V_* / k)\left(\ln \frac{z}{z_0}\right)\left[1 - \frac{\psi_m\left(\frac{z}{L}\right)}{\ln\left(\frac{z}{z_0}\right)}\right] \quad \text{(A5)}$$

Donelan (1982) indicated the relationship between the stability parameter $\psi_m(z/L)$ and the Richardson number (Ri):

For unstable conditions (Twater > Tair) $\psi_m(z/L) = 7.6$ Ri
For stable conditions (Twater < Tair) (z/L) = 6.0 Ri

The Richardson number being defined as,

$$Ri = \frac{gz(T_{air} - T_{water})}{(T_{air} + 273.15)V_z^2} \quad \text{(A6)}$$

For, respectively, stable and unstable conditions, Hsu et al. (1999) proposed,

$$\psi_m\left(\frac{z}{L}\right) = -5\frac{z}{L} \quad \text{(A7)}$$

$$\psi_m\left(\frac{z}{L}\right) = 1{,}0496\left(-\frac{z}{L}\right)^{0{,}459} \quad \text{(A8)}$$

Because of local particularities, which reflect much of the weather as roughness, several authors have indicated the importance of theoretical development of time and specific value for the parameterizations discussed above. As example, note the diurnal variability in the relation $_{Tair} - _{Twater}$ for tropical reservoir, that reach up to 5°C during day and -6°C during night (Figure A1). Applying this at Equation A6, results in a diurnal variability for Ri and, consequently, variable conditions for the stability of the lower PBL.

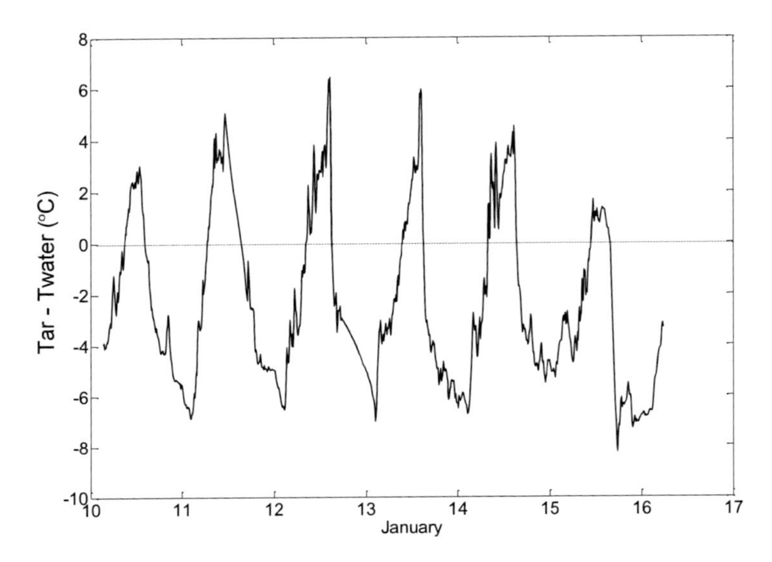

Figure A1.Difference between air and surface water temperatures at Manso reservoir for a week of January.

References

Alcântara, E.; Novo, E.; Stech, J.; Assireu, A.; Nascimento, R.; Lorenzzetti, J.; Souza, A. Integrating historical tographic maps and SRTM data to derive the bathymetry of a tropical reservoir. *Journal of Hydrology*, 389, 311-316, 2010.

Amarante, O.A.C.; Zack, M.B.J.; Leite de Sá, A. *Atlas do Potencial Eólico Brasileiro.* Brasília, Brasil, Centro de Pesquisas de Energia Elétrica (Cepel), 2001.

ANEEL. Banco de Informações de Geração. Agência Nacional de Energia Elétrica. Brasília, Brasil: ANELL; 2010.

Archer, C.L.; Jacobson, M.Z. Evaluation of global wind power. *J. Geophys Res.* 2005; 110, B22110, doi:10.1029/2004JD005462.

Bitencourt, D.P.; Acevedo, O.C.; Moraes, O.L.L.; Degrazia, G.A. A interação do vento local no interior de um vale com o escoamento de grande escala – análise de dois estudos de caso. *Rev. Bras. Meteor.*, 24(4), 436-447, 2009.

Booth J.G., Miller, R.L.; McKee, B.A.; Leathers, R.A. Wind-induced bottom sediment resuspension in a microtidal coastal environment. *Conti. Shelf Res.* 20, 785-806, 2000.

Bullard, J. E.; Wiggs, G.F.S.; Nahs, D. J. Experimental study of wind directional variability in the vicinity of a model valley. *Geomorphology*, 35, 127-143, 2000.

Farr, T. G., et al. The Shuttle Radar Topography Mission, *Rev. Geophys.*, 45, 2007. doi:10.1029/2005RG000183.

Fearnside, P.M. Greenhouse gas emissions from hydroelectric dams: controversies provide a springboard for rethinking a supposedly 'clean' energy source - an editorial comment. *Climate Change* 66:1–8, (2004).

Feitosa, E.A.N.; Pereira, A.L.; Silva, G.R.; Veleda, D.R.A.; Silva, C.C. Panorama do potencial eólico do Brasil. Brasília, Brazil: Dupligráfica, 2003.

Finnigan, J. J.; Raupach, M.R.; Bradley, E.F.; Aldis, G.K. A wind tunnel study of turbulent flow over a two-dimensional ridge. *Boundary-Layer Meteorol.* 50, 277–317, 1990.

Garcia, A.; Torres, J. L.; Prieto, E. and Francisco, A. Fitting wind speed distributions: A case study. *Solar Energy*, 6 (2), 139-144, 1998.

Garvine R. and Kempton W. Assessing the wind field over the continental shelf as a resource for electric power. *Journal of Marine Research*, 66 (6), 751-773, 2008.

Guardans, R.; Palomino, I. Description of wind field dynamic patterns in a valley and their relation to mesoscale and synoptic-scale meteorological situations. *J. Appl. Meteorol.*, 34, 49–67, 1995.

GWEC. Global Wind 2009 Report. Global Energy Council, GWEC, 56 pp. 2009.

Hess, G. D. and Garratt J. R. Evaluating models of the neutral, barotropic planetary boundary layer using integral measures: Part I. Overview. Boundary-Layer Met., 104, 333–358, 2002.

Hsu, S. A. Estimating overwater friction velocity and exponent of Power-Law wind profile from gust factor during storms. *J. Waterway and Oc. Engineering*, 174-177, 2003.

Hunt, J. C. R.; Leibovich, S.; Richards, K.J.; Turbulent shear flows over low hills. *Quart. J. Roy. Meteo. Soc.*, 114, 1435–1470, 1988a.

Hunt, J.C.R.; Richards, K.J.; Brighton, P.W.M. Stably stratified shear flow over low hills. *Quart. J. Roy. Meteo. Soc.*, 114, 859–886, 1988b.

Huthcance, J. M. Wind-driven circulation in coastal and marginal seas. *Can. J. Remote Sensing*, 28(3), 329-339, 2002.

IPCC. Intergovernmental Panel on Climate Change. Climate Change: The Physical Science Basis. Summary for Policymakers. Working Group I of the IPCC, Paris, February 2007.

Justus, C. G. and Mikhail, A. Height variation of wind speed and wind distribution. *Geophys. Res. Lett.*, 3, 261-264, 1976.

Justus, C.G.; Hargreaves, W. R.; Mikhail, A.; Graber, D.; Methods for estimating wind speed frequency distributions. *J. Appl. Met.*; 17, 350-353, 1978.

Koçak, K. A method for determination of wind speed persistence and its application. *Energy*, 27, 967-973, 2002.

Lucena, A. F. P.; Schaeffer, R and Szklo, A. S. Least-cost adaptation options for global climate change impacts on the Brazilian electric power system. *Global Environment Change*, 20, 342-350, 2010.

Lucena, A. F. P.; Szklo, A. S.; Schaeffer, R and Dutra, R. M. The vulnerability of wind power to climate change in Brazil. *Renewable Energy*, 35, 904-912, 2010.

Manwell, J. F.; McGowan J. G. and Rogers, A. L. Wind energy explained, John Willey and Sons, West Sussex, England, 590 pp., 2002.

Mason, P.J. Flow over the summit of an isolated hill. *Boundary-Layer Meteorol.* 37, 385–405, 1986.

Massey Jr, F. J. The Kolmogorov-Smirnov test for goodness of fit. *J. Amer. Statist. Ass.* 46 (70) 1951.

Niemer E. Climatologia do Brasil, 2nd edtition. IBGE, 421 pp., 1989.

Nfaqui, H.; Buret, J.; Sayigh, A. A. M. Wind characteristics and wind energy potential in Marrocos. *Solar Energy*, 6(1), 51-60, 1998.

Offer, Z.Y.; Goossens, D. The use of topographic scale models in predicting eolian dust erosion in hilly areas: field verification of a wind tunnel experiment. *Catena*, 22, 249–263, 1994.

Pimenta, F. M.; Kempton, W.; Garvine, R. Combining meteorological stations and satellite data to evaluate the offshore wind power resource of Southeastern Brazil. *Renewable Energy*, 33, 2375-2387, 2008.

Rehman, S.; Halawani, T. O. and Hussain, T. Weibull parameters for wind speed distribution in Saudi Arabia. *Solar Energy*, 3(6), 473-479, 1994.

Sansigolo, C. A. Distribuições de probabilidade de velocidade e potência do vento. *Rev. Bras. Meteor.*, 20 (2), 207-214, 2005.

Silva, B. B.; Alves, J. J., Cavalcanti, E. P. Caracterização do potencial eólico da direção predominante do vento no estado da Bahia. In : XII *Congresso Brasileiro de Meteorologia,* 2002.

Stevenson, M.R.; Lorenzzetti, J.A.; Stech, J.L.; Arlino, P.R.A. SIMA - An Integrated Environmental Monitoring System. In: VII Simpósio Brasileiro de Sensoriamento Remoto, Curitiba, 1993. Anais. Curitiba: INPE, 1993. 4: 300-310, 1993.

Van der Auwera, L.; Meyer, F. and Malet, L. M. The use of the Weibull 3-parameters model for estimating mean wind power densities. *Journal of Applied Meteorology*, 19 (7), 819-825, 1980.

Weigel, A. P.; Rotach, M. W. Flow structure and turbulence characteristics of the daytime atmosphere in a steep and narrow Alpine valley. Quart. *J. Roy. Meteor. Soc*., 130 (602), 2605-2627, 2004.

Whiteman, C.D.; Doran, J.C. The relationship between overlying synoptic-scale flows and winds within a valley. *J. Appl. Meteorol.* 32, 1669–1682, 1993.

Zoumakis, N. M. The dependence of the Power-law exponent on surface roughness and stability in a neutrally and stably stratified surface boundary layer. *Atmósfera*, 6, 79-83, 1993.

In: Energy Resources
Editor: Enner Herenio de Alcantara
ISBN: 978-1-61324-520-0

Chapter 2

BIOENERGY IN CENTRAL EUROPE – RECENT DEVELOPMENTS, INTERNATIONAL BIOFUEL TRADE AND FUTURE PROSPECTS

Gerald Kalt*[*], *Lukas Kranzl and Reinhard Haas
Energy Economics Group, Vienna University of Technology,
Gusshausstrasse, Vienna, Austria

ABSTRACT

In order to assess future prospects of bioenergy use, it is essential to have thorough knowledge of the status quo, recent developments and unused primary energy potentials.

To this end, statistical data on the current biomass use and international biomass streams as well as data on biomass potentials in literature need to be reviewed and discussed. In this chapter, this is done for the Central European (CE) region, with a special focus on international biomass trade and the situation in Austria.

The contribution of biomass and wastes to the energy supply in CE countries ranges from 2.8% in Italy to 14.9% in Denmark (2008). Due to European directives and according national support schemes, the share of biomass in the total energy consumption increased significantly in recent years, especially in Denmark (+6% from 2000 to 2008), Germany (+4.8%), Austria (+4.5%) and Hungary (+3.9%). The main progress was achieved in

[*]E-mail: kalt@eeg.tuwien.ac.at; Tel.: +43 1 58801 370363; Fax: +43 1 58801 370397.

the field of electricity and CHP generation as well as the production and use of transport fuels.

With regard to the compilation and interpretation of statistics on internationally traded biomass volumes, various challenges need to be addressed.

Statistical data on cross-border trade often do not cover the whole range of biomass used for energy recovery, such as energy crops for biofuel production or biomass which is intended for material uses and ultimately end up in energy production.

Therefore, methodological approaches to gain insight into recent developments and the status quo of biofuel trade are proposed and discussed. Subsequently, it is analysed which Central European countries act as importers and exporters of biomass, and trade streams are mapped.

The main importers of wood fuels in CE are Italy, Denmark and Austria. Cross-border trade of wood pellets has increased significantly in recent years. For Denmark pellets are the most important biomass import stream. Austria, being a net exporter of wood pellets, is importing considerable amounts of wood residues, primarily indirectly in the form of industrial roundwood.

With regard to direct biofuel trade (biodiesel and ethanol), Austria, Italy and Poland are the main importers (primarily biodiesel). Although growing rapidly, cross-border trade related to biofuels for transport is still rather moderate compared to (indirect and direct) trade of wood fuels in CE.

However, there is strong evidence that the CE region is currently becoming increasingly dependent on imports of biofuels as well as feedstock for biofuel production.

For the case of Austria, a detailed assessment of trade streams, including trade streams which are not considered in energy statistics, namely indirect trade of wood-based fuels and energy crops intended for biofuel production is carried out.

The results indicate that the net imports of biomass accounted for up to on fourth of the total bioenergy use in Austria in recent years. This is about three times the quantity that energy statistics suggest.

The results and methodological approaches of studies assessing biomass potentials indicate that there are considerable unused biomass resources available in Austria and other CE countries.

This chapter also provides insight into the (among the considered countries highly inhomogeneous) structure and current exploitation of biomass potentials and the achievable contribution to the total energy supply.

Finally, conclusions about future prospects for bioenergy use with a focus on the EU's 2020 targets and policy recommendations are derived.

1. INTRODUCTION AND OUTLINE

Among the different renewable energy sources (RES) bioenergy is of crucial importance for the current and future energy supply in Central Europe (CE).[1] Not only because it already has the highest share of all RES, but also due to the vast potentials of biomass and the fact that it can be used in all energy sectors: for heat-only, electricity or combined heat and power generation as well as for the production of transport fuels.

With regard to the "2020-RES-targets" (as defined in the 2009-EU Directive on the promotion of the use of energy from renewable sources; EC, 2009a) the current structure of bioenergy use, recent developments and the availability of environmentally compatible resource potentials are of high interest.

Within this chapter recent developments are discussed, implications with regard to cross-border trade and energy policy targets are analysed. The considered countries include Austria, Czech Republic, Germany, Hungary, Poland, Slovenia and Slovakia as well as Italy and Denmark[2]. The sections of this chapter are organized as follows:

After this introduction, section 2 provides insight into the structure of energy consumption and bioenergy use in CE (sections 2.1. to 2.2.3). A special focus is given to the situation in Austria (section 2.3). The topic of section 3 is international trade of biomass.

After discussing methodological issues of assessing cross-border trade related to bioenergy (3.1), net imports and exports according to energy statistics are analysed (3.2). In section 3.3 trade streams of wood in CE are mapped. For the case of Austria a more detailed assessment of cross-border trade related to bioenergy use is carried out (3.4).

Section 4 deals with prospects for a further increase of bioenergy use, EU energy policy framework conditions and biomass resource potentials in the considered countries. Section 9 includes a discussion, conclusions and policy implications.

[1]Within this chapter, "bioenergy" is used for all kinds of biomass utilization for energy recovery. "Biofuels" comprise liquid and gaseous biogenic fuels used for transportation, such as biodiesel and bioethanol.

[2]These countries are referred to as "CE countries" in this chapter, even though Italy and Denmark are usually not considered to be part of Central Europe. They have been included primarily because of their significant cross-border trade streams as well as their characteristic biomass consumption profiles.

2. THE CONTRIBUTION OF BIOMASS TO THE ENERGY SUPPLY

2.1. The Structure of Energy Consumption in Central Europe

Despite the geographical vicinity of the considered countries, the structures of their primary energy consumption (gross inland consumption; GIC) are quite inhomogeneous (Figure 1). On an average the share of fossil fuels (petroleum, natural gas, lignite and hard coal) accounts for 80% of the total energy sources used, with Slovenia and Slovakia being least dependent on fossil fuels (both about 70%). The share of hard coal and lignite ranges from less than 10% (Italy) to more than 50% (Poland) and the contribution of petroleum from 21% (Slovakia) to 43% (Italy). The share of natural gas is especially high in Hungary's and Italy's GIC (both close to 40%) and relatively low in Poland and Slovenia (both slightly more than 10%). In the Slovak Republic nuclear energy accounts for as much as 23% of the GIC, whereas in Austria, Denmark, Italy and Poland there are no nuclear power plants in operation. There are also significant differences with regard to energy consumption per capita. In Hungary and Poland it accounts for 108 GJ/a, whereas in the Czech Republic it is 182 GJ/a and in Germany 175 GJ/a. In the other countries it ranges from about 125 to 170 GJ/a. (All data refer to 2008.)

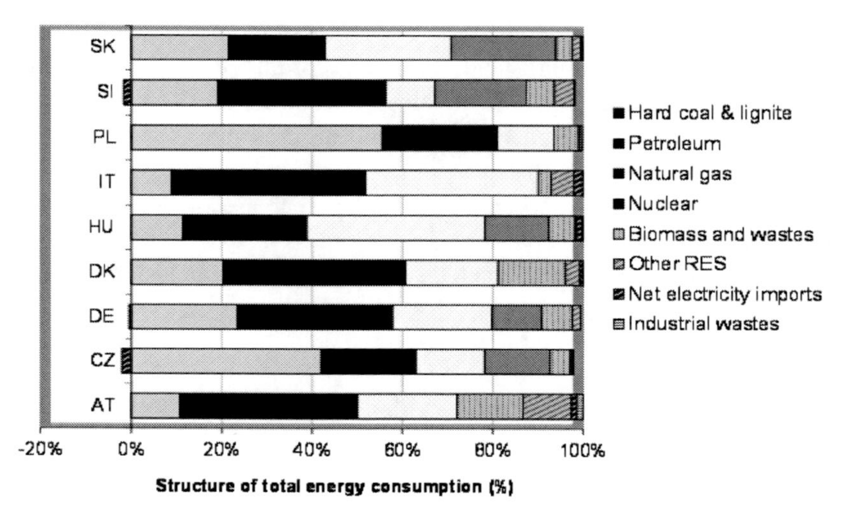

Source: Eurostat (2010a), own calculations.

Figure 1. Structure of the GIC in CE countries in 2008.

2.2. The Contribution of Bioenergy in Central Europe

The shares of renewable energies in the GIC of the considered countries range from 5% in the Czech Republic to 25.3% in Austria (2008), with biomass and wastes accounting for an average of more than 70% of all renewables.[3] In the Czech Republic, Poland and Hungary biomass and wastes account for more than 90% of all RES. The share of biomass and wastes in the GIC is illustrated in the map in Figure 2. It is highest in Denmark (14.9%) and Austria (14.7%). The high contribution in Denmark is a result of ambitious energy policy measures which led to a significant increase of biomass use in combined heat and power (CHP) and district heating plants (largely based on imported biomass, as will be shown in section 3), especially since the early nineties. For the case of Austria the following reasons for the high importance of biomass have been identified: (i) Austria is a heavily wooded country. Almost 50% of the total Austrian area is forests, which is clearly more than in most other CE countries.[4] (ii) The use of biomass for residential heating is traditionally high in Austria. Especially in the eighties log wood boilers gained in importance due to the oil price shock and in recent years pellet boilers and other modern biomass heating systems have become increasingly popular (partly due to attractive investment subsidies).

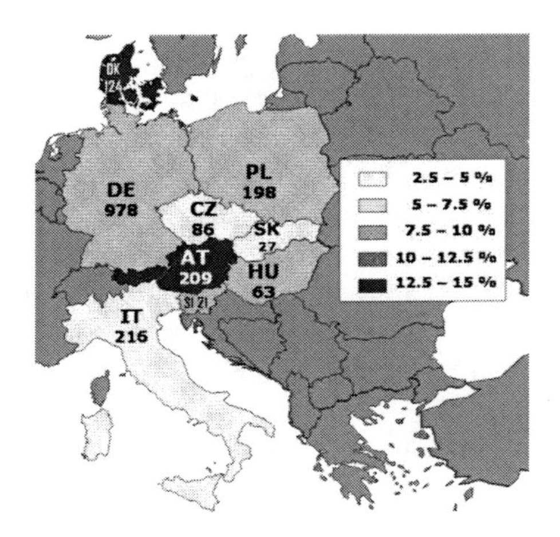

Source: Eurostat (2010a).

Figure 2. Bioenergy as share of gross inland energy consumption in 2008 (values in PJ/a).

---

[3]The fact that non-renewable wastes are also included in "biomass and wastes" is neglected here.

[4]Only Slovenia has an even higher share of approximately 60%.

Today more than 20% of the total residential heat demand is met with biomass (Statistik Austria, 2010b). (iii) The prominent role of the wood processing industries in Austria was crucial for the development of the bioenergy sector. First, they provide substantial amounts of wood residues for energy use and second, a high proportion of their energy demand is covered with biomass. Therefore, the bioenergy share in the energy supply of the industrial sector is also exceptionally high.

2.2.1. The Structure of Biomass Use and Recent Developments

In Figure 3. the historic development of the share of biomass and wastes in the total GIC is illustrated for each CE country. The figure shows that in most countries the contribution of biomass increased significantly in recent years. The most notable developments were achieved in Germany and Denmark, but also in Austria, Czech Republic, Hungary and Slovakia the importance of biomass for energy production has been increasing steadily; especially since the year 2000 or so. In Austria the biomass consumption has more than doubled from 1990 to 2008, but due to the rising total energy consumption (about 34% increase from 1990 to 2008), the biomass share only showed an increase of about 60%.[5] In absolute numbers the biomass consumption in CE increased from about 450 PJ in 1990 to 1,950 PJ in 2008. Remarkably, the progress in Germany accounted for more than 50% of this increase. In 2008 about 50% of the total amount of biomass used for energy recovery in CE was consumed in Germany. The biomass consumption per capita is highest in Austria (25 GJ in 2008), followed by Denmark (22.7 GJ), Germany (12 GJ) Slovenia (10.5 GJ).

Figure 4. shows that the main increase in bioenergy use was achieved in the field of electricity and CHP generation. The share of biomass for heat-only production, accounting for about 80% in the nineties has recently gone down to less than 50%. The main reason for the increase in electricity and CHP generation was the implementation of the "EC Directive on electricity production from renewable energy sources" (EC, 2001) and the introduction of according support schemes (e.g. the German Renewable Energy Sources Act).

Among the considered countries the ratio of electricity generation from biomass and wastes to the total electricity consumption ranges from less than 1% in Slovenia to more than 10% in Denmark. In Austria (6.4%), Germany (5.3%) and Hungary (4.7%) the ratio is also relatively high, whereas in Czech Republic, Italy, Poland and Slovakia it is only about 2%. In the early nineties only the biomass share in Austria's electricity consumption accounted for more than 2%.

[5] In most CE countries the energy consumption declined during this period.

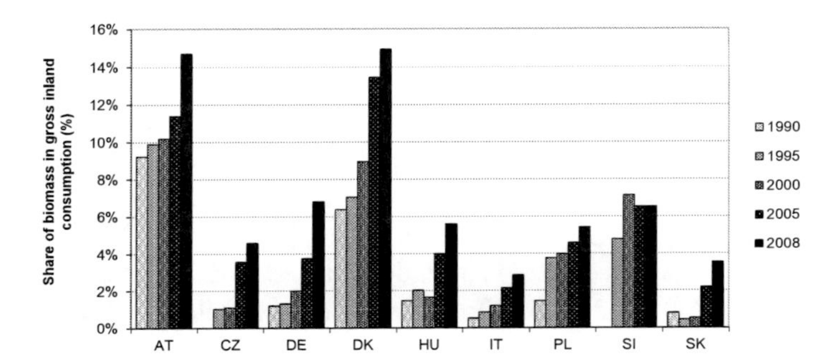

Figure 3. Development of bioenergy as share of GIC from 1990 to 2008.
Source: Eurostat (2010a), own calculations.

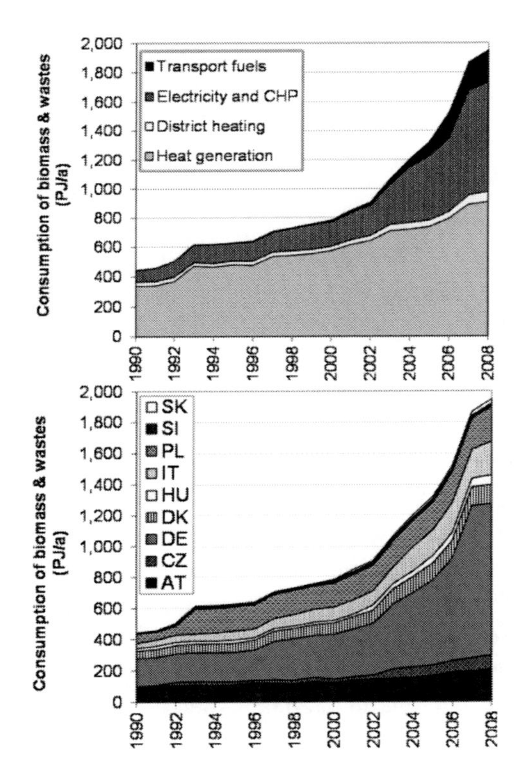

Source: Eurostat (2010a), own calculations.

Figure 4. Biomass consumption in CE countries from 1990 to 2007 broken down by application (left) and country (right).

2.2.2. Biofuels for Transport

The main progress in the use of biofuels started in 2003, as a consequence of Directive COM 2003/30/EC on the promotion of the use of biofuels for transport ("Biofuel Directive"; EC, 2003). According to the directive, EU Member States are required to establish national targets on the proportion of biofuels in the transport sector.

The following reference values for national targets are stated in this directive: 2% by the end of 2005 and 5.75% by the end of 2010, calculated on the basis of energy values.

The progress in the considered countries according to the national progress reports in the context of the Biofuel Directive (EC, 2009b and EC, 2010a) as well as the national target values are illustrated in Figure 5. The figure illustrates that there are sometimes significant differences between the data according to Eurostat (2010a) (represented by error bars) and the data stated in the biofuel reports, indicating that the consumption of biofuels is partly not captured appropriately in energy statistics. This is particularly true for Slovakia as well as for the 2008-data for Italy and Poland.

However, progress was very uneven among CE countries. Based on the national progress reports, Austria had the highest share of biofuels in 2009 (7%), followed by Germany (5.5%), Poland (4.63%), Hungary (3.75%) and Italy (3.47%). Up to 2007, Germany was the European leader in the field of biofuels. It had already surpassed its 2010-target of 6.25% in 2006, but in 2009 the share of biofuels had dropped to 5.5% due to an abolishment of the tax exemption for biofuels (see section 2.2.3).

In most other CE countries no appreciable progress was reported until 2008 or 2009. Denmark's latest report (for the year 2008) indicates a biofuel share of only 0.12%. According to DEA (2009), Denmark aims at achieving the indicative 5.75%-target in 2012, after a gradual phase-in starting in 2010.

Figure 6 shows the historic development of biodiesel and bioethanol production in CE countries. Germany is the major producer of both biofuels. The German biodiesel production accounted for about 50% of the total production in the EU in the years 2002 to 2007.

Thereafter the production in Germany declined and its share in the total production in the EU decreased to about 28% (2009).[6]

---

[6]An increase in tax levels for pure biodiesel in Germany in 2007 has severely affected the competitiveness of biodiesel, and numerous production plants have gone out of operation.

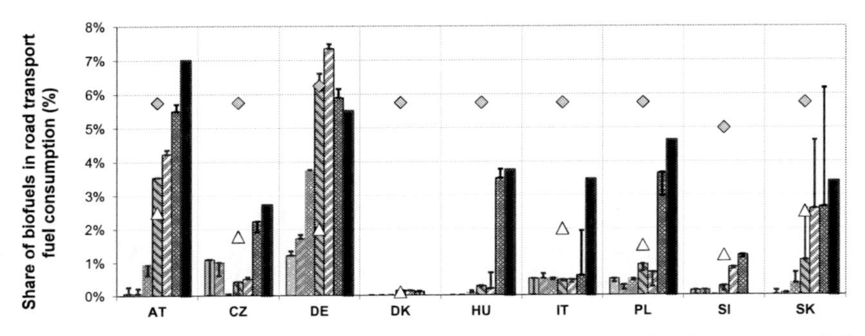

Source: EC (2009b), EC (2010a) (no data for 2009 available for Denmark and Slovenia); error bars: data according to Eurostat (2010a), own calculations.

Figure 5. Share of biofuels for transport and national indicative target values in the context of Directive COM 2003/30/EC.

The capacity of biodiesel and bioethanol production plants being built recently in CE is considerable: From 2007 to mid-2010, the installed biodiesel production capacities increased from 3.8 Million tons per year (Mt/a) to 9.7 Mt/a (EBB, 2011). The bioethanol production capacities installed in CE increased from 1.94 Mt/a in mid-2008 to 3.1 Mt/a in mid-2010 (ePURE, 2011). About 50% of these capacities are located in Poland (0.56 Mt/a).

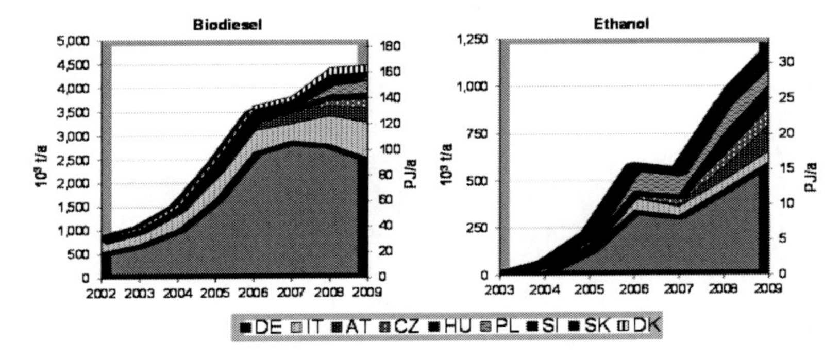

Source: EBB (2011), ePURE (2011), own calculations[7].

Figure 6. Production of biodiesel and bioethanol in CE countries (quantity in tons and net calorific value of the fuels produced).

[7]For biodiesel production only aggregated data for Denmark and Sweden are available. The data for Denmark shown in the figure are therefore based on the installed production capacities.

At full capacity, biodiesel and bioethanol plants installed in mid-2010 could produce as much as 7.8% of the total fuel consumption in road transport in CE (2008). Hence, with regard to the available production capacities, the 5.75%-target for 2010 could theoretically be easily achieved. However, actual production figures have been clearly below production capacities and the question of whether or not the target will be achieved remains questionable; especially with regard to the recent developments in Germany. (Throughout the EU-27 the indicative target is very unlikely to be reached according to Resch et al., 2008a).

According to EC (2009b) the self-sufficiency of biofuels for transport of the EU-27 (defined as the ratio of production to consumption) decreased from 109% in 2005 to 73% in 2007. Throughout CE countries, the self-sufficiency of biodiesel was 83% and the one of bioethanol 76% in 2009 (calculation based on EBB, 2011, ePURE, 2011 and preliminary data according to EurObserv'ER, 2010). However, as biofuels are partly produced with imported feedstock, these calculations actually do not bear any information as to what extent the biofuel supply is based on imports. This aspect will be discussed in more detail for Austria and Germany in section 3.

2.2.3. Support Schemes for Bioenergy

In the field of transport and electricity generation from RES, EU directives issued shortly after 2000 resulted in a notable growth in bioenergy use in most CE countries. For heat generation no such directive was issued before Directive 2009/28/EC ("2009-RES-Directive"; EC, 2009a) and policy support was limited to diverse national or regional support schemes. These include investment subsidies (e.g. Austria, Germany, Slovenia), tax incentives (e.g. Austria, Germany), bonuses to electricity feed-in tariffs for the utilization of waste heat from combined heat and power plants (e.g. Czech Republic, Germany), certificate systems (e.g. Italy) and soft loans (e.g. Poland, Slovenia) (Resch et al., 2008b).

The most common instruments to promote biofuels in the transport sector are tax relieves and obligations to blend. According to EC (2009b) all CE countries used tax exemptions as the main support measure in 2005 and 2006. In Austria and Slovakia there were also obligations to blend. Since 2007 this policy instrument has also been adopted in Germany, Czech Republic, Italy and Slovenia, mostly in combination with increasing levels of taxation. For example in Germany the law on biofuel quotas ("Biokraftstoffquotengesetz") which came into force in January 2007 put an end to total tax exemption and established an obligation to blend (4.4% for biodiesel in diesel fuel and 1.2% for bioethanol in petrol).

A major indirect support scheme for bioenergy and other low-carbon RES is the EU Emission Trading Scheme for greenhouse gases (EU ETS), which operates in the EU-27 plus Iceland, Liechtenstein and Norway. It was launched in 2005 and covers CO2 emissions from power stations, combustion plants and other industrial plants with a net heat excess of more than 20 MW (EC, 2010b).

2.3. The Development of Bioenergy Use in Austria

This section provides a more detailed insight into the historic development of biomass use in Austria, based on national statistics which are more detailed than the ones available on Eurostat (2010a). Figure 7 shows the development of biomass primary energy consumption broken down by biomass types. From 1970 to 2004, biomass statistics differentiated only between the categories "wood log" and "other biomass and biofuels". The data for the biogenic fraction of municipal solid wastes are estimates based on the total energy use of wastes and an assumed biogenic share of 20%. More detailed data are available for the years 2005 to 2009, as shown in the figure. The biogenic share of wastes was in the range of 17 to 24% during this period.

Figure 7 also shows the share of biomass in the total gross inland consumption, which increased from less than 6% (less than 50 PJ/a) during the mid-1970 to 15% (210 PJ) in 2009. The main increase in biomass use took place during the periods 1980 to 1985 and 2005 to 2009. Until the year 1999 the use of wood log for domestic heating accounted for more than 50% of the total biomass use for energy. The rest was primarily wood wastes and residues of the wood processing industries as well as waste liquor of the paper and pulp industry. Especially during the last five years, the different fractions of wood biomass, including forest wood chips, industrial residues and other wood wastes as well as liquid and gaseous biomass have become increasingly important, whereas wood log remained relatively constant at about 60 PJ/a. Hence, wood log accounted for only 30% of the total biomass use in 2009.

Figure 8 shows the development of biomass final energy consumption from 1970 to 2009. The data are broken down by fuels used for residential heating or industrial heat production (further broken down by wood log and other biomass), district heat, electricity and transport fuels produced from biomass.[8] In 2009, wood log and other biogenic fuels used for heat generation accounted for 65.6%

[8]"Final energy consumption" covers energy supplied to the final consumer for all energy uses.

of the biomass final energy consumption, district heat generated with biomass for 13.5%, electrical energy from biomass power plants for 8.5%, and transport fuels for 12.4%.

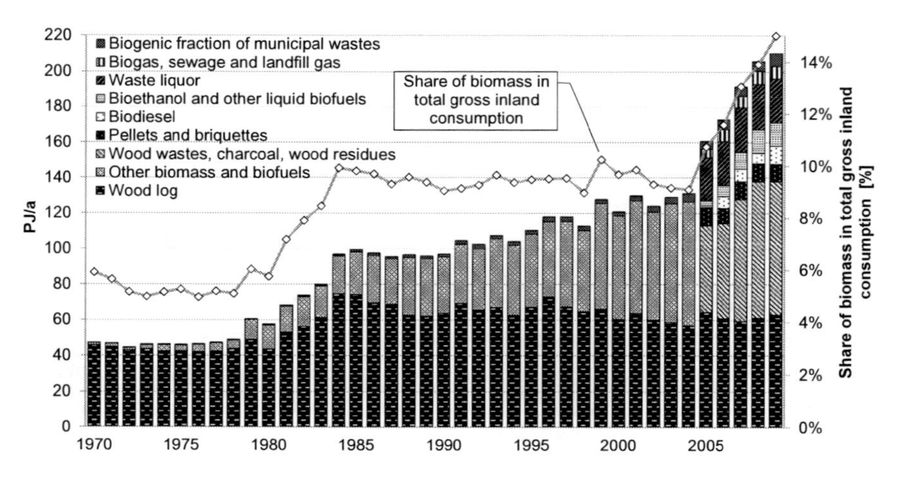

Source: Statistik Austria (2010a).

Figure 7. Biomass gross inland consumption in Austria from 1970 to 2009 and biomass share in total primary energy consumption.

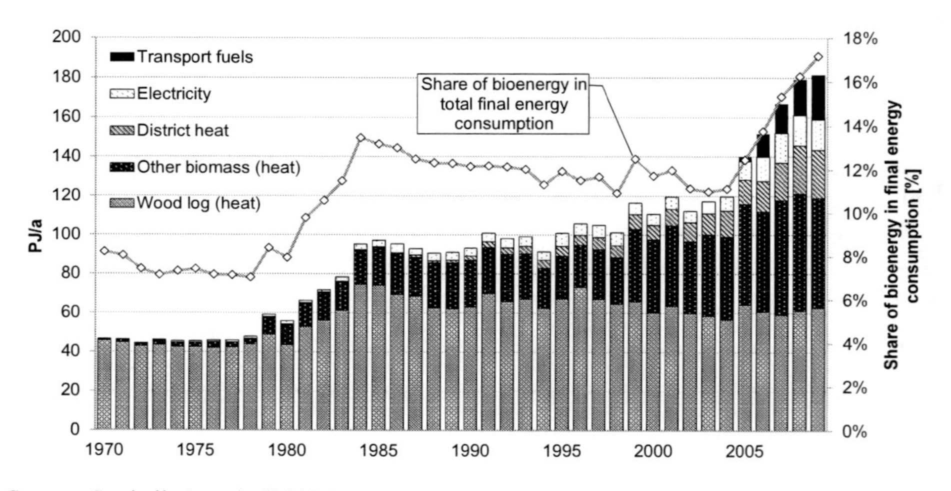

Source: Statistik Austria (2010a).

Figure 8. Biomass final energy consumption in Austria from 1970 to 2009 and biomass share in total final energy consumption.

3. CROSS-BORDER TRADE OF BIOMASS FOR ENERGY

Well-functioning international biomass markets are considered one of the key factors for mobilizing the global biomass production potential and serving the growing demand for biomass for energy (Heinimö et al., 2007). Heinimö and Junginger (2009) argue that international biomass trade for energy is still in its initial phase and global trade volumes of certain biomass types (e.g. wood pellets, ethanol or plant oil) have already increased significantly in recent years.

Projects studying international bioenergy markets and trade have been launched, such as the international collaboration project entitled "Task 40: Sustainable International Bioenergy Trade: Securing Supply and Demand", which is carried out within the framework of the IEA Bioenergy agreement (see IEA, 2010a and IEA, 2010b). The objective of Task 40 is to "support the development of a sustainable, international, bioenergy market, recognising the diversity in resources [and] biomass applications [...] by providing high quality information and analyses for market players, policy makers, international bodies as well as NGOs".

More specifically, one of the core objectives is to "map and provide an integral overview of biomass markets and trade on global level". The analyses presented in this section are intended to contribute to this objective by providing insight in current state of cross-border trade in CE, the impact of increasing bioenergy use on biomass streams as well as by carrying out a critical review of data in statistics and discussing methodological aspects.

3.1. Methodological Aspects

As Heinimö and Junginger (2009) emphasize, no comprehensive statistics and summaries aggregating separate biomass trade flows for energy generation are available and there are several challenges related to measurement of internationally traded volumes of biomass for energy generation.

Many biomass streams are traded for several applications, including both material and energy purposes (e.g. wood chips or oilseeds and plant oil for the production of biodiesel) or they are traded for material uses and ultimately end up in energy production (indirect trade).

Feedstock used for biofuel production is generally not taken into account energy statistics. Table 3 1 gives an overview of the methodological approaches applied in this section, their advantages and drawbacks as well as the biomass types considered and references/databases used. For data from trade statistics the CN codes of the respective commodities are provided (see EC, 2007).

The following methodological approaches are applied: First, the net imports (or net exports) of the following biomass types are analysed on the basis of energy and other statistics[9] (section 7.2): wood and wood waste, wood pellets, biodiesel and bioethanol (direct trade) and wood residues (indirect trade in the form of roundwood in the rough).

These data provide a rough overview about which countries act as net importers and exporters, and on the importance of direct cross-border trade of biomass for energy.

Next, direct trade streams of fuelwood and other wood fractions in the CE region are mapped, in order to identify the main trade streams of wood fuels (section 7.3).

Table 1. Data used and methodologies applied for assessing biomass trade

Short description	Types of biomass, databases/references used, CN codes	Characteristics and features (favourable: +, adverse: –)
Assessment of net imports / net exports based on energy and other statistics	Wood and wood wastes used for energy (Eurostat, 2010a) Wood pellets (Pellet@las, 2010) Indirect trade of wood residues (based on roundwood statistics according to FAO, 2010a) Biodiesel and bioethanol (Eurostat, 2010)	Avoidance of error sources related to trade statistics Trade streams of products with no separate CN codes can be assessed Volumes which are not covered in trade statistics can be assessed (e.g. blends of biofuels with fossil fuels) Neglect of the lag between production and consumption as well as stockkeeping results in errors No information about trade partners Trade of upstream products (e.g. energy crops for transport fuel production) is not taken into account
Short description	Types of biomass, databases/references used, CN codes	Characteristics and features (favourable: +, adverse: –)

[9]Apart from energy statistics (Eurostat, 2010a), data have been obtained from Pellet@las (2010) and FAO (2010a).

Investigation and mapping of trade statistics	Wood residues (UN Comtrade, 2009; CN codes 4401 2100, 4401 2200 and 4401 3010) Fuelwood (UN Comtrade, 2009; CN code 4401 1000)	Use of official data on international trade volumes Information about trade partners available Several error sources related to trade statistics, e.g. shipments below declaration limit not included, commodities may be recorded under wrong CN Codes, country of origin or ultimate destination may be unknown in case of transit No differentiation between energy and non-energy use (no separate CN Codes) Several biomass types sometimes aggregated under one CN Code (e.g. pellets included in wood residues) Only quantities of specific products included; trade of upstream products not considered (e.g. trade of oilseeds or plant oil intended for biodiesel production)
Assessment of total cross-border trade related to bioenergy use (exemplary assessment for the case of Austria)	Direct trade: energy statistics (Eurostat, 2010a) Indirect trade with wood-based fuels: statistics of wood processing industries and supply statistics (BMLFUW, 2010, FAO, 2010a etc.) Biofuel production and consumption statistics (Winter, 2010), supply balances for agricultural commodities (Statistik Austria, 2010c)	Provides comprehensive insight into biomass trade relevant for bioenergy use Indirect trade streams and trade with upstream products (feedstock for biofuel production) can be assessed High data requirements, data need to be collected from different databases and statistics of industries Complete assessment of indirect trade streams not possible due to insufficient data availability Preselection of commodities is necessary; selection is not straightforward and background knowledge of trade streams is required

Table 1. (Continued)

Assessment of direct and indirect effects of biomass use on trade flows of related products	Biodiesel (impact on oilseed and plant oil trade streams);Databases: EBB (2009); rapeseed production: Eurostat (2010a), UN Comtrade (2009); CN codes 1205, 1514, 1511	Suitable for fuels with several upstream products which can be used for energy and other purposes Indirect and spillover effects can be assessed Selection of commodities which are taken into account is not straightforward; background knowledge/presumptions on indirect effects required Only rough conclusions are possible due to uncertainties related to other influencing factors Information on conversion processes required

However, since the statistical data used for these approaches are fragmentary and do not cover the whole range of biomass trade relevant for bioenergy use, a complete assessment is carried out for the exemplary case of Austria (section 7.4). This includes an assessment of indirect trade of wood-based fuels and of feedstock used for biofuel production. The assessment of indirect trade is based on a comprehensive analysis of international and domestic wood trade flows (i.e. imports and exports of the wood-processing industries as well as trade streams between the industries), in order to capture the total amount of biomass used for energy generation and originating from non-domestic production. Finally, the impact of increasing resource demand for biodiesel production on trade statistics of oilseeds and plant oil is exemplarily analysed for the cases of Germany and Austria (section 7.5).[10]

3.2. Net Imports and Exports of Biomass

Under disregard of the time lag between biomass production and consumption, the difference can be considered as net imports (or net exports, respectively).

[10]For the conversion of trade data given in mass units to energy units, the following LHV are assumed: Fuelwood and wood residues: 14.4 MJ/kg, wood pellets: 18 MJ/kg, biodiesel: 37 MJ/kg, bioethanol: 26.7 MJ/kg.

The main advantage of this simple approach is that the numerous error sources related to trade statistics are avoided. Apart from errors caused by neglecting stockkeeping, which can be especially relevant during very dynamic market developments, the main drawback is that no information about trade partners can be obtained.

3.2.1. Wood and Wood Waste

Figure 9. shows the net imports of "wood and wood wastes", based on energy statistics (Eurostat, 2010a).[11] The data indicate that especially the net imports of Italy and Denmark have increased significantly in recent years.

More than 30% of the wood biomass consumption in Italy and about 25% of the consumption in Denmark is based on imports. According to ENS (2009), the net imports of wood chips, wood pellets and fuelwood accounted for 19.5 PJ in 2008 (1.8, 15.5 and 2.2 PJ, respectively).

Austria has turned from net exporter to net importer in recent years, reflecting the increasing demand for wood fuels during this period. Czech Republic on the other hand has been exporting increasing amounts of wood biomass.

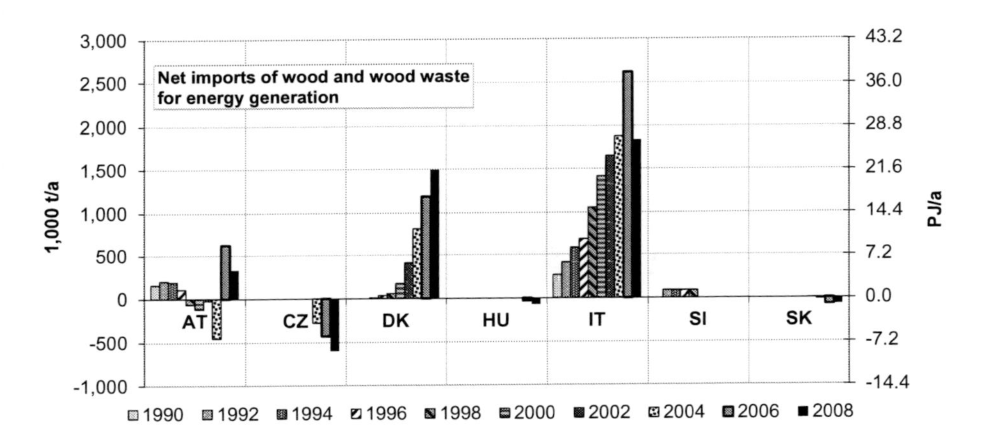

Source: Eurostat (2010a); own calculations.

Figure 9. Net imports of wood biomass for energy generation based on energy statistics (no data for Germany and Poland available).

[11]According to the definition by Eurostat, the category "wood and wood wastes" covers "a multitude of woody materials generated by industrial processes or provided directly by forestry and agriculture (firewood, wood chips, bark, sawdust, shavings, chips, black liquor, etc.) as well as wastes such as straw, rice husks, [...] and purpose-grown energy crops (poplar, willow, etc.)".

3.2.2. Wood Pellets

Wood pellets are well suited for transportation due to their high density and energy content. Recent policy and market changes have stimulated an increasing demand for wood pellets (Peksa-Blanchard et al., 2007) and given an impetus to international trade with wood pellets. Figure 10 illustrates the net imports of wood pellets from 2001 to 2008. The increase in international trade is especially apparent in the data for Austria, Germany, Denmark and Poland. Denmark and Italy have been importing significant amounts of wood pellets in recent years, whereas the other CE countries are net exporters. It is remarkable that pellets account for the largest single fraction of wood fuels imports to Denmark. With regard to net exports it has to be mentioned that the neglect of the time lag between production and consumption may result in an overestimation.

3.2.3. Indirect Imports of Wood Residues

A large percentage of roundwood material in the rough being shipped for the purpose of sawnwood production actually ends up as byproducts (bark, sawdust, wood chips etc.). Due to the vast amounts of roundwood being traded globally, these indirect imports of wood residues are of some significance. Heinimö and Junginger (2009) conclude that indirect trade of biomass through trading of industrial roundwood and material byproducts composes the largest share of global biomass trade.

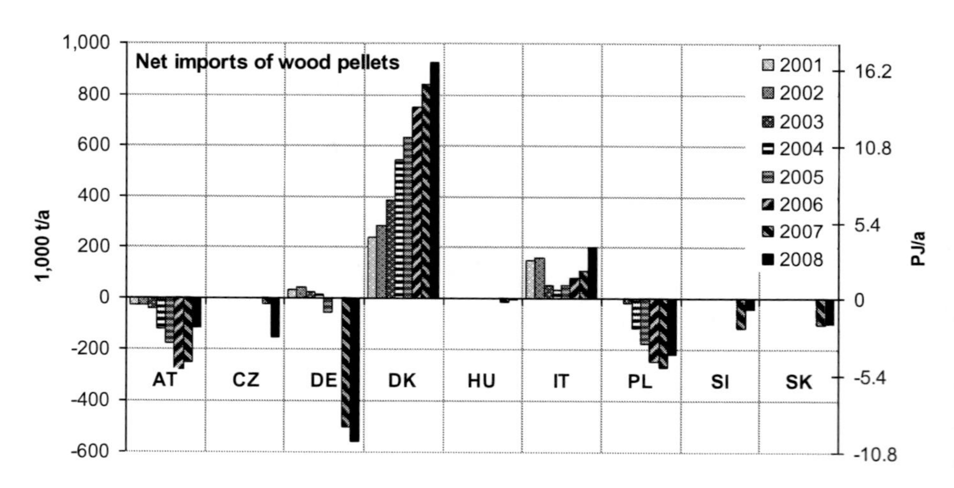

Source: Pellet@las (2010), own calculations.

Figure 10. Net imports of wood pellets based on production and consumption statistics.

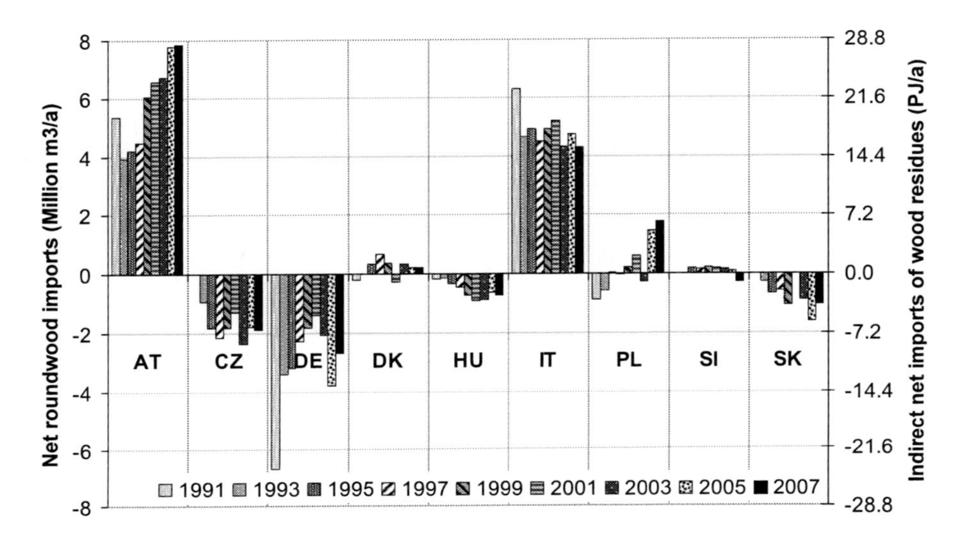

Source: FAO (2010a); own calculations.

Figure 11. Net imports of roundwood (in million m3; left axis) and the according indirect net imports of wood residues (in PJ; right axis)[12].

Figure 11. shows the net imports of roundwood and the estimated indirect net imports of wood residues from 1991 to 2007.[13] Austria and Italy are the main importers of industrial roundwood in CE. While Italy shows a declining trend, Austria's net imports have almost doubled since the mid-1990s. The main exporters of industrial roundwood are Germany and the Czech Republic.

Figure 11. provides a rough overview into the quantities of indirectly traded wood residues, and into which countries are net importers and which are net exporters of roundwood.

However, it needs to be considered that wood residues are not only used for energy recovery but also for material uses, primarily the production of paper, pulp and wood boards. Therefore it is necessary to analyse the trade flows within the countries, in order to gain insight into the quantities relevant for bioenergy use. In section 3.4.3 this is done for the case of Austria.

---

[12]Based on Heinimö and Junginger (2009) who estimate that 40–60% of roundwood can be converted into forest products, it is assumed that 50% of the industrial roundwood end up as residues.

[13]There are several other streams of indirect biomass imports, including for example waste wood in the form of wood products or residues from sawnwood processing. However, cross-border trade of industrial roundwood is assumed to be by far the most significant indirect biofuel stream.

3.2.4. *Liquid Biofuels for Transport*

With the growing demand for biofuels for transport[14], the volumes of internationally traded biofuels have been increasing strongly in recent years. The total biodiesel imports of CE countries (with trade between CE countries included) increased from 70,000 t in 2005 to about 800,000 t in 2008 and the total exports from 210,000 t to 480,000 t. The total bioethanol imports increased from zero to 420,000 t and the exports from 30,000 t to 190,000 t during the period 2005 to 2008 (Eurostat, 2010a).

The following figures show the development of net imports of biodiesel and ethanol for the CE countries as well as the aggregated data for the CE region. Apparently, Austria was the main net importer of biodiesel in the considered period, whereas Czech Republic, Germany and Denmark stand out as net exporters.

The net trade flows of bioethanol are much lower, except for the case of Poland. With regard to the development during the period 2005 to 2008, which was characterized by substantial increase in biofuel use in CE (see section 2.2.2), it is evident that production could not keep pace with the growing demand. The aggregated data for all considered countries illustrate that despite the rapidly increasing production (see Figure 6) the CE region turned from a net exporter into a net importer of both biodiesel and ethanol.

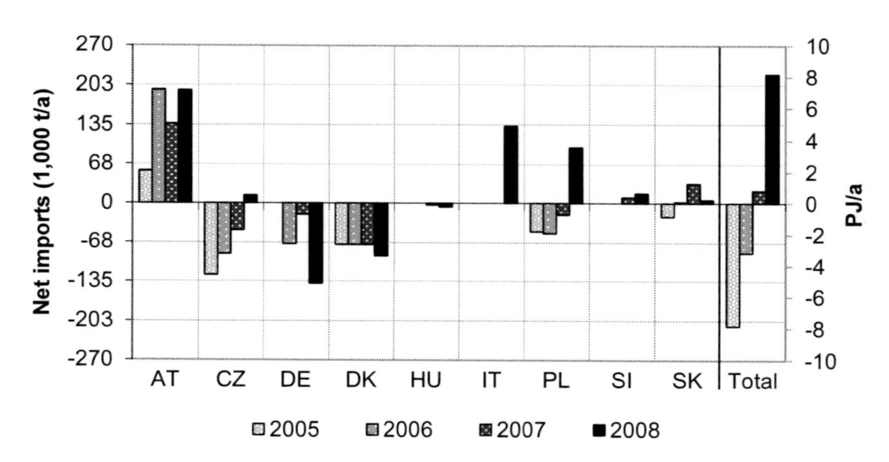

Source: Eurostat (2010a), own calculations.

Figure 12. Net imports of biodiesel based on energy statistics.

[14]Only biodiesel and bioethanol are considered here. Apart from these biofuels, vegetable oil is of some significance in Germany and Austria (EurObserv'ER, 2010).

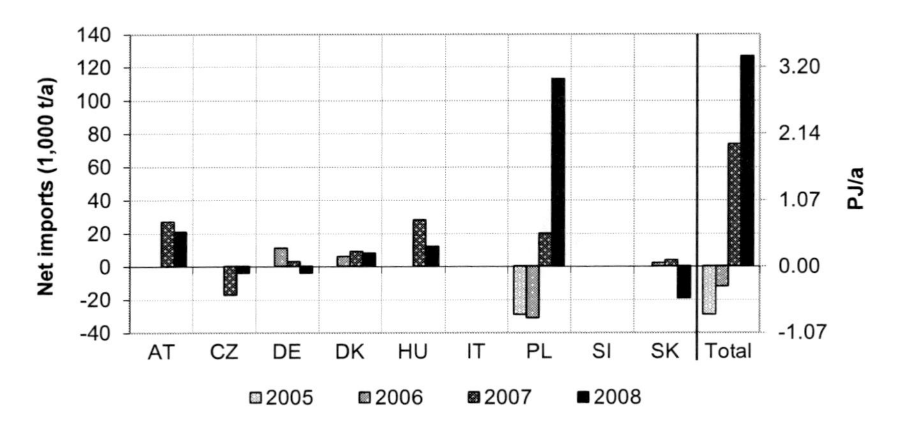

Source: Eurostat (2010a), own calculations.

Figure 13. Net imports of bioethanol based on energy statistics.

Analyses of biofuel trade streams (based on UN Comtrade, 2009, for example) prove to be problematic, as statistical compilation of biofuel trade of biofuels for transport is still in the early stages. Only since January 2008, there is a separate CN Code for biodiesel (3824 9091). Before that date, biodiesel had to be classified under a general CN subheading together with other chemical products[15] (Freshfields Bruckhaus Deringer, 2008). Furthermore, the quantities reported under the newly established CN Code are highly incomplete, as only biodiesel shipped in its pure form is included.[16]

Bioethanol is classified under CN code 2207 0000, together with any other sort of "denatured ethyl alcohol and other spirits of any strength", making it impossible to map trade streams of bioethanol used as transport fuel. Apart from that, like biodiesel ethanol is also shipped in blends of different proportions, further complicating analyses of trade streams.

3.3. Streams of Wood Biomass in Central Europe

Based on trade statistics (UN Comtrade, 2009), the following figures show wood biomass streams in the CE region. Figure 14. shows the trade streams of

[15]CN Code 3824 9098 "chemical products and preparations of the chemical or allied industries, including those consisting of mixtures of natural products".

[16]For example, the biodiesel imports reported by Austria in 2008 account for only 20% of the import quantities according to energy statistics. These incomplete data suggest that Austria imported more than 80% of the total imports from Germany.

fuelwood (CN code 4401 1000 “wood in logs, in billets, in twigs, in faggots or in similar forms”) in the year 2007. With total net imports amounting to 7.4 PJ in 2007, Italy is the main importer of fuelwood. However, Italy’s fuelwood imports in 2007 accounted for only slightly more than 20% of its total imports of wood biomass (cp. Figure 9). More than 50% of Italy’s fuelwood imports come from CE countries. The rest is imported primarily from Croatia and Bosnia-Herzegovina. Further major fuelwood streams are from the Netherlands (i.e. from overseas) to Germany and from Ukraine to Hungary. Austria is also importing noteworthy amounts of fuelwood from Czech Republic, Slovakia and Hungary.

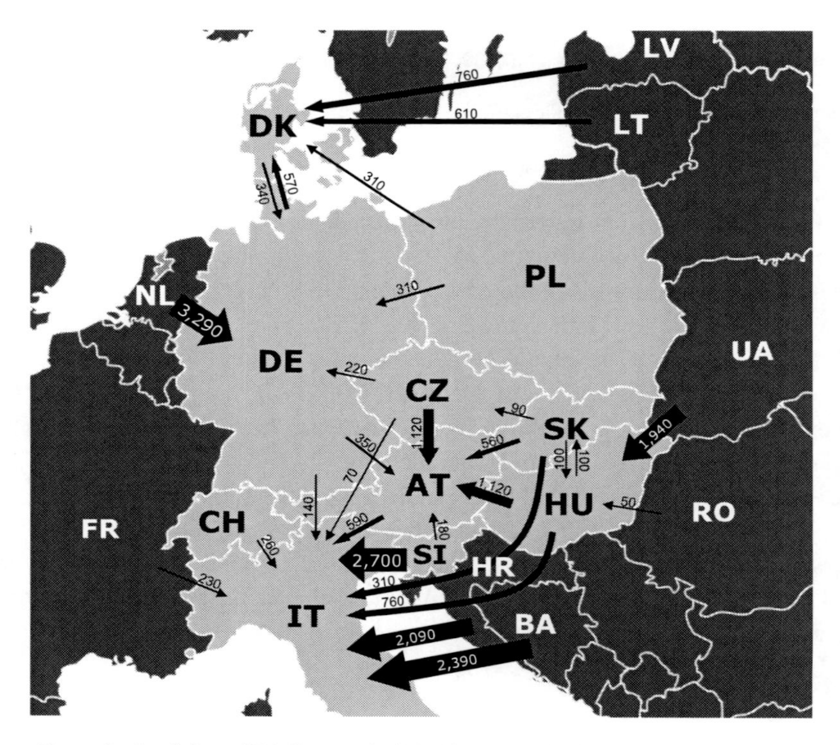

Source: Data obtained from UN Comtrade (2009), own calculations and illustration.

Figure 14. Cross-border trade of fuelwood in Central Europe in 2007 (in TJ/a; flows smaller than 50 TJ/a are not depicted; unlabeled neighbouring countries do not have any relevant trade flows)[17].

[17]Data reported by the importing and the exporting country often show significant discrepancies; in Figure 14 and Figure 15 always the higher value is shown.

However, in total the net imports to Austria accounted for less than 5% of its total fuelwood consumption in 2007. It has to be noted that the data reported in trade statistics are connected with high uncertainties. This becomes obvious when data reported by the importing country are compared with the respective data reported by the exporting country, which are often highly inconsistent. It is assumed that these discrepancies are due to different regulations concerning the notification of imports and exports, as well as methodologies of data collection. Even though fuelwood cross-border trade among some CE countries has been increasing substantially in recent years (especially imports to Italy, increasing by close to 400% in the last ten years or so according to UN Comtrade, 2010), it is concluded that the trade volumes of fuelwood are rather insignificant in relation to its utilization in CE.

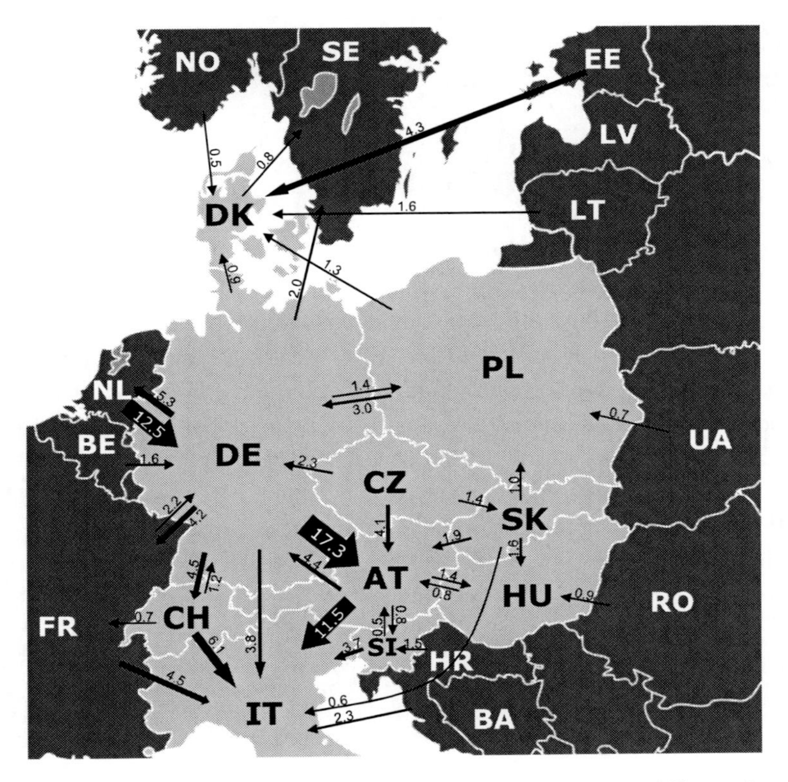

Source: Data obtained from UN Comtrade (2009), own calculations and illustration.

Figure 15. Cross-border trade of wood residues (including wood chips, sawdust, briquettes, pellets etc. for energy and material purposes) in Central Europe in 2007 (in PJ/a; flows smaller than 0.5 PJ/a are not depicted).

Figure 15. illustrates the cross-border trade of wood chips, sawdust, pellets etc. (in the following the term "wood residues" is used for these fractions)[18]. The quantities are clearly higher than those of fuelwood shown above.

However, this category also includes wood which is used for material purposes. It is clear to see that apart from German overseas imports via the Netherlands and Denmark's imports from the Baltic States, the main streams are Austria's imports from Germany and Austria's exports to Italy.

The figures confirm that Austria and Italy are the main net importers of wood residues in CE. For the case of Austria, this is partly due to the high demand of the paper and pulp industry and the board industry, but the share of wood residues being used for energy generation has been increasing significantly in recent years (see section 3.4.3).

From 1996 to 2007 Austria's total import quantity of wood residues increased from 0.85 Mt to 1.9 Mt. However, clearly larger volumes are imported indirectly through industrial roundwood, as it was shown in section 3.2.3.

3.4. Cross-Border Trade Related to Bioenergy Use in Austria

This section provides a more detailed insight into the importance of biomass cross-border trade for bioenergy use in Austria. Apart from energy statistics (Statistik Austria, 2010a), production and consumption statistics of the wood-processing industries (sawmill industry: FAO (2010a), paper and pulp industry: Austropapier (2010), wood board industry: Schmied (2009), statistical data on wood consumption and trade (FAO, 2010b), supply balances for agricultural commodities (Statistik Austria, 2010c) as well as reports on timber felling (BMLFUW, 2010) and biofuel consumption (Winter, 2010) were used. Hence, the data required for gaining insight into the importance of international trade streams for the bioenergy sector go far beyond energy statistics provided by Eurostat or national statistical institutes, respectively.

3.4.1. Biomass Trade According to Energy Statistics

Figure 16. shows the imports and exports of biomass used for energy production in Austria according to energy statistics, broken down by the different

[18] "Wood in chips or particles": sawdust and wood waste and scrap, whether or not agglomerated in logs, briquettes, pellets or similar forms (CN Codes 4401 2100, 4401 2200 and 4401 3010).

types of biofuels, pellets and briquettes, wood log and charcoal.[19] Primarily due to the relatively high imports of wood log and biodiesel, the net imports were clearly positive since 2006. In the years 2006 and 2009 they accounted for close to 10% of the GIC of biomass in Austria.

3.4.2. Cross-Border Trade of Biofuels

The increasing use of biogenic transport fuels (biodiesel, vegetable oil and ethanol) in recent years resulted in a significant increase of cross-border trade. Apart from direct trade with biofuels, cross-border trade of feedstock used for biofuel production need to be taken into account.

Biodiesel

Figure 17. shows the development of biodiesel production and direct imports and exports according to the official biofuel reports persuant Directive 2003/30/EC (Winter, 2010). The figure shows that imports accounted for approximately 50% of the inland consumption in the period 2005 to 2009. Close to one fourth of the domestic production of biodiesel, which increased from 70,000 t (2005) to more than 320,000 t (2009) during this period was exported.

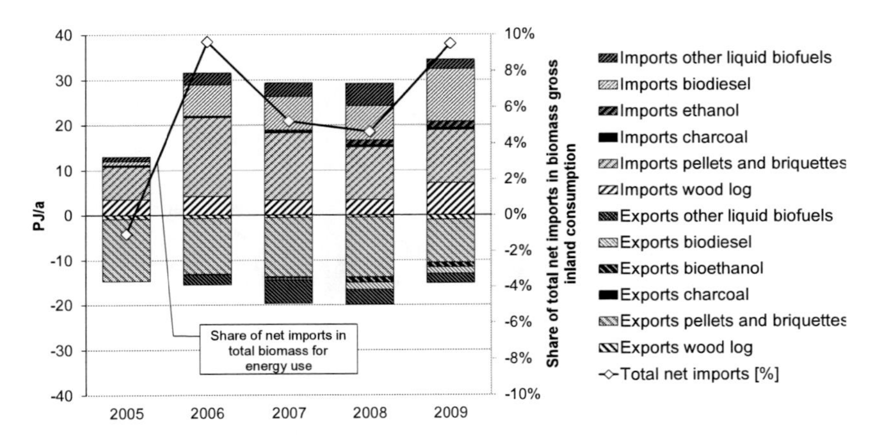

Source: Statistik Austria (2010a), own calculations

Figure 16. Imports and exports of biogenic energy carriers according to energy statistics.

[19]These detailed data are only available for the period 2005 to 2009 (cp. Figure 7 and Figure 8). A comparison with data presented in the previous sections indicates that the category "pellets and briquettes" also includes unrefined wood residues.

With regard to plant oil used for transportation, there are hardly any reliable data, as production volumes in statistics are not differentiated by intended uses and due to largely regional distribution channels. According to Winter (2010), approximately 17,000 to 18,000 t (0.6 to 0.67 PJ) of plant oil were used for transportation annually during 2007 to 2009. It is assumed that at least the quantities which are used in agriculture (approximately 2,700 t or 0.1 PJ in the year 2009) originate from domestic production.

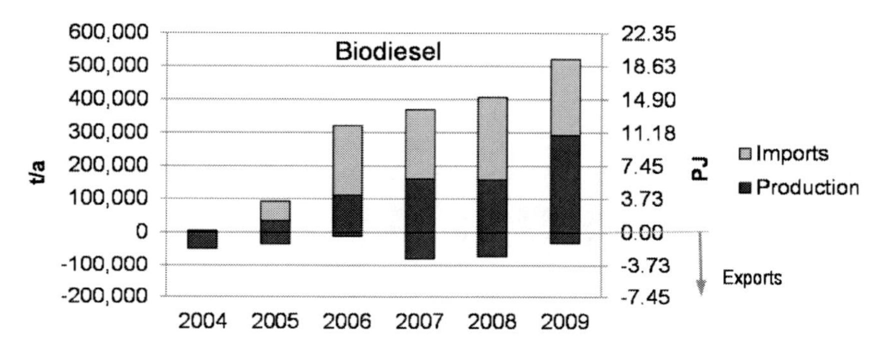

Source: Winter (2010), own calculations.

Figure 17. Austrian biodiesel supply from 2004 to 2009 according to the official biofuel report persuant Directive 2003/30/EC (Stockkeeping is neglected).

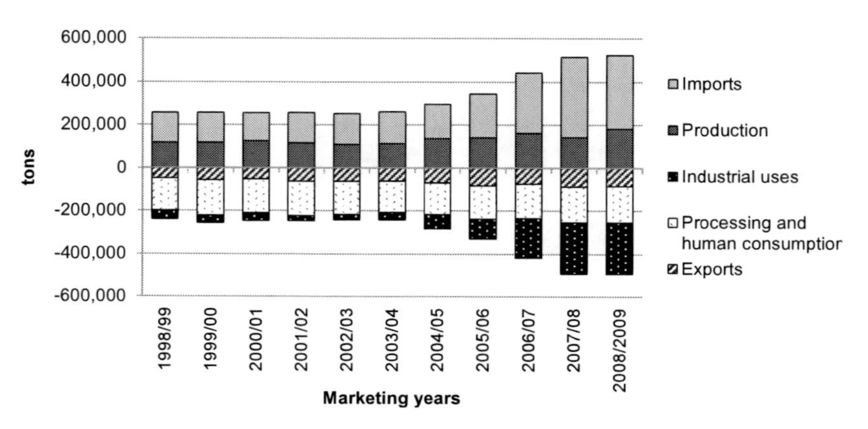

Source: Statistik Austria (2010c).

Figure 18. Supply balance for vegetable fats and oils (losses, stockkeeping and animal feed are not shown due to negligible quantities).

In order to provide insight into the impact of biodiesel and plant oil for energy use on Austria's trade streams, the supply balance for vegetable fats and oils is shown in Figure 18. The supply balance shows "sources" (imports and domestic production) as well as "sinks" (processing and human consumption, exports and industrial uses). It is clear to see that the rapidly increasing industrial use of vegetable oils and fats (i.e. primarily biodiesel production) was facilitated by a significant increase in imports, whereas domestic production remained relatively constant. The self-sufficiency (calculated on the basis of the oil yield from domestic oilseed production) decreased from about 60% (marketing years 1998/99 to 2000/01) to less than 30% (2007/08: 23%, 2008/09: 27%). Today, industrial uses exceed the quantity used for processing and human consumption in Austria.

To conclude, the additional demand for energetic uses of vegetable fats and oils was almost exclusively covered with imports. The most important trade streams are rapeseed oil imports from the eastern neighboring countries and Eastern Europe, respectively, but Austria is also importing increasing amounts of palm oil: From 2000 to 2008 the net imports increased from about 13,000 t to 47,000 t (UN Comtrade, 2009).

Bioethanol

The Austrian production of bioethanol used for transportation is limited to one large-scale plant, located in Pischelsdorf in Lower Austria and operated by the AGRANA holding company. The plant became fully operational in mid-2008 (in 2007 a test run was carried out) and has a capacity of approximately 190,000 t/a (5.1 PJ/a). Figure 19 shows the bioethanol production, imports and exports in Austria from 2007 to 2009. Whereas in 2007 and 2008, Austria was a net importer of bioethanol, the net exports in 2009 amounted to about 28% of the production. The annual feedstock demand at full capacity is reported to account for 620,000 t (75% wheat and triticale, 15% maize and 10% sugar juice). According to Kopetz et al. (2010), the agricultural land used for the production of "ethanol feedstock" in 2007 was 6.749 ha. There are no profound data available on the feedstock supply in 2008 and 2009, but according to the operator's financial report for the business year 2009/10 (AGRANA, 2010), most originated from domestic production.

The self-sufficiency of cereals varied from 94 to 110% in the marketing years 2003/04 to 2008/09. Despite the additional demand for ethanol production (about 400,000 t), the self-sufficiency in 2008/09 was as high as 105%, because the production quantity in this marketing year surpassed the average of the previous five years by about 1 Mt.

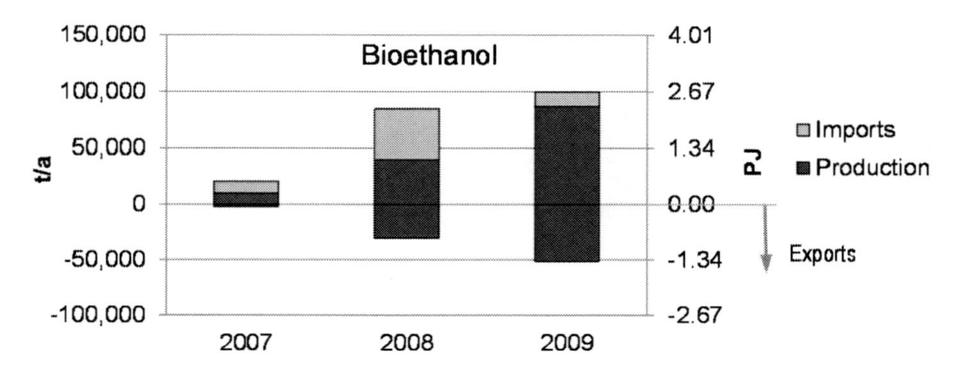

Source: Winter (2010), own calculations.

Figure 19. Austrian bioethanol supply from 2007 to 2009 according to the official biofuel report persuant Directive 2003/30/EC.

Hence, it is concluded that (i) the feedstock demand for bioethanol production is relatively moderate, compared to the total cereal production (approximately 5.75 Mt in 2008/09), (ii) based on historical data, general conclusions about the impact of bioethanol production on international trade streams in Austria are not possible, but (iii) the data for 2008/09 suggest that the feedstock demand for the current quantity of bioethanol production can basically be supplied from domestic production without reducing the self-sufficiency.

3.4.3. Indirect Cross-Border Trade of Wood-Based Fuels

As mentioned before, indirect imports of biomass include quantities which are originally imported for material uses but ultimately end up in energy generation. For the case of Austria, the presumably most significant indirect trade streams of wood-based fuels are indirect imports of residues of the sawmill industry (industrial wood residues), bark from industrial roundwood, residues of the wood board industry and waste liquor of the paper and pulp industry.

In order to assess the indirect biomass imports used for energy, it is essential to have an idea of the different utilization paths of the various wood fractions, as well as the trade streams between the wood processing industries: The bulk of industrial roundwood is processed to sawnwood by the sawmill industry. The average share of imported roundwood in the consumption of the sawmill industry was about 45% in the last ten years. The paper and pulp industry as well as the wood board industry process roundwood and wood residues of the sawmill industry. Therefore, the sawmill industry acts as an important raw material supplier for the other industry segments. The increasing production of the Austrian sawmill industry in the last years and decades provided favourable framework conditions for the growth of the paper and pulp and the wood board industry. However, in recent years the demand for wood residues for energy generation has been increasing significantly, and the import quantities of these industries segments have also amounted to notable trade streams.

Based on production and consumption statistics of the wood processing industries as well as trade statistics, the quantities of the indirect trade streams mentioned above have been assessed: During the period 2001 to 2009 imports accounted for an average of 42% (between 35 and 52%) of the wood consumption of the Austrian sawmill industry and the share of sawmill residues being used energetically increased from 12% in 2002 to about 40% in 2009 (own calculations based on statistical data of the wood processing industries). Accordingly, the energy quantity of indirectly imported sawmill residues increased from 2.7 PJ in 2002 to 8.1 PJ in 2009. Futhermore, the quantity of bark being imported in the form of roundwood and used for energy production is estimated 5.8 PJ/a (average value of the period 2001 to 2008; no discernible trend during this period). With regard to waste liquor of the wood processing industries, the analysis of statistical data indicates that between 38 and 44% of the total quantity reported in energy statistics can be traced back to imports (directly imported roundwood and wood residues as well as indirectly imported residues). Hence, on an average about 10 PJ of indirectly imported waste liquor were used for energy production annually during 2001 to 2009. Compared to this, the quantities of indirectly imported wood residues of the board industries are relatively low (about 2.6 PJ/a).

The results of the assessment of indirect imports of wood-based fuels in Austria is summarized in Figure 20. In total, indirect imports of wood-based fuels amounted to an energy equivalent of more than 25 PJ/a since 2006. At the same time, the share of indirect imports in the total biomass gross inland consumption in Austria has declined from more than 18% (2002 and 2004) to around 13% (2008 and 2009), as the total biomass consumption has increased more rapidly than the contribution of indirect imports.[20]

Due to the economic crisis in 2009, a downturn of the production quantities of all wood-processing industries could be observed.

[20]The annual fluctuations of indirect imports are partly due to weather conditions and storms, which had a significant impact on the domestic wood supply in recent years (e.g. the storms "Kyrill" and "Paula" in 2007 and 2008, respectively).

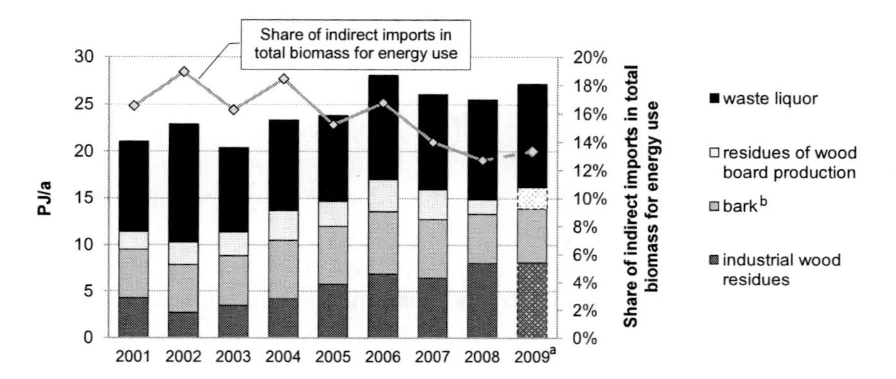

Source: FAO (2010a), FAO (2010b), Austropapier (2010), BMLFUW (2010), Hagauer (2007), Schmied (2009), own calculations.

Figure 20. Development of indirect imports used for energy, and the according share in the total biomass consumption in Austria. a) The data for indirect imports of industrial wood residues and residues of wood board production in 2009 are estimates, as no 2009-data on the wood consumption of the board industry were available at the time these analyses were carried out.b) The share of bark in imported quantities of industrial roundwood is assumed 10%.

However, the relative decrease of sawnwood production (and therefore also the inland supply of industrial residues) decreased more significantly (minus 24%) than the production of the paper and pulp and the board industry (minus 12% and minus 11%, respectively). The paper and pulp industry's consumption of domestically produced industrial residues decreased by more than 30%[21], and the share of imports in the total wood consumption increased to about 30% (compared to an average share of 22% in the previous five years). It is important to note that there are some other indirect biomass trade streams, which are not taken into account here: First of all, streams of wood products like sawnwood, wood panels, paper etc. which usually end up in energy generation, either in dedicated bioenergy plants utilizing waste wood, or in waste treatment plants. There are substantial methodological challenges related to the assessment of these indirect trade streams, including insufficient statistical data on trade volumes, uncertainties about the lifetime of wood products, recycling rates and many more. With regard to wood products Austria a net exporter, which puts the high indirect imports shown in Figure 20 somewhat into perspective.

[21]For the estimated 2009-data in Figure 20 it was assumed that the same shift in the wood consumption structure occurred in the board industry.

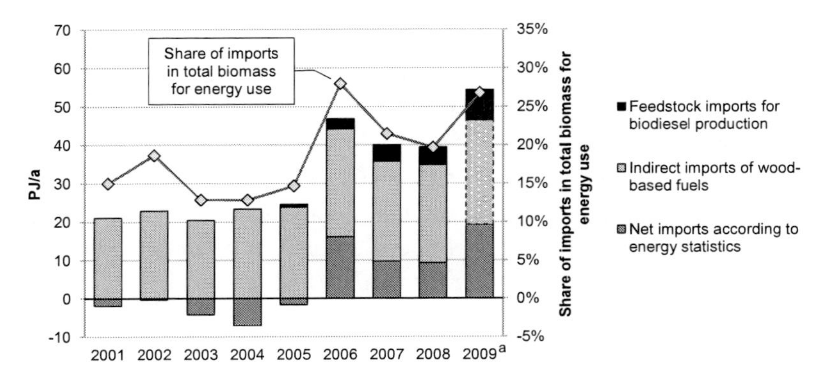

Figure 21. Development of total biomass imports for energy use including indirect imports and feedstock for biodiesel production, and the according share in the total biomass use in Austria, a) The data for indirect imports of wood-based fuels in 2009 are partly based on estimates.

3.4.4. Summary

To sum up, with feedstock for biofuel production and indirect trade streams taken into account, cross-border trade of biomass for energy use is clearly more significant than energy statistics suggest. Based on the assessments described above, it is concluded that the share of imported biomass was between 20% and 30% of the total biomass consumption in Austria during 2006 to 2009 (Figure 21). Indirect imports of wood-based fuels are the most significant fraction, but direct imports of wood fuels, liquid transport fuels and feedstock imports for biofuel production have also become increasingly important in recent years.

3.5. The Impact of Biodiesel Production on International Trade Streams

A crucial issue in connection with the increasing demand for biofuels are possible indirect effects and spillover effects, especially on global land use and food markets. There has been growing concern about possible impacts of biofuel production from edible crops on global food security as well as sustainability issues like indirect land-use change (see EC, 2010c).

For example Fischer et al. (2010a) argue that "uncoordinated biofuels development can contribute substantially to short-term price shocks [...] and may also result in a stable trend in rising food prices".

Plant oil and oilseeds are basically more suitable for long-distance transportation than wood biomass due to their higher specific calorific values. In section 3.2.4 it was shown that the rising consumption of biofuels was accompanied by increasing direct cross-border trade.

The objective of this section is to analyse the impact of the increasing biodiesel production on trade streams of plant oil and oilseeds. The focus is on the countries which showed the most rapid development in biodiesel production and consumption among CE countries: Germany and Austria.

The basic approach is to convert production data of oilseeds, net imports of oilseeds and plant oil and data on the demand for biodiesel production on a common basis of comparison ("plant oil equivalents") and to qualitatively investigate correlations between the time series (see Table 3 1 for details on the commodities considered, references and CN Codes). Figure 22. and Figure 23. show the data for Germany and Austria, respectively. It is clear to see that in both countries, the growing plant oil demand for biodiesel production primarily resulted in an increase in rape oil net imports, rather than domestic production of rapeseeds. Furthermore, in both countries also a (compared to the rape oil imports moderate but still notable) increase in palm oil imports, primarily from Indonesia and Colombia, took place. Germany's total palm oil net imports increased from 0.44 Mt in 2000 to about 0.9 Mt in 2008 and Austria's net imports from 16,000 to 47,000 t during the same period.

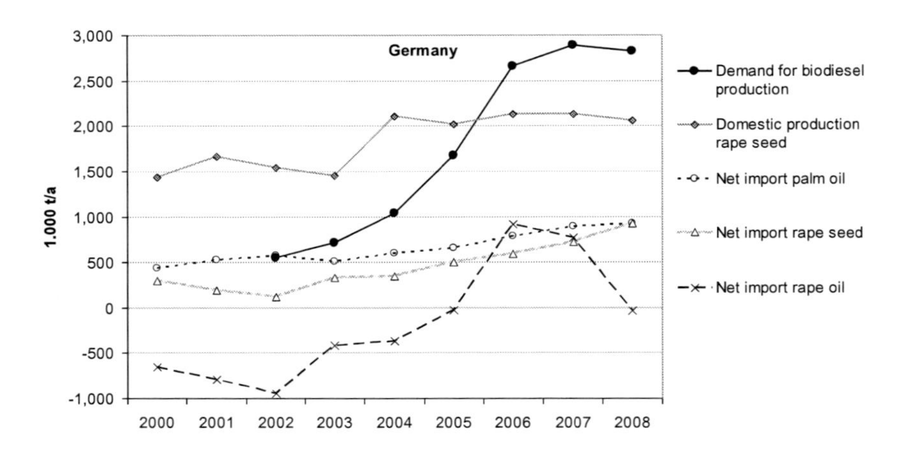

Source: UN Comtrade (2009), Eurostat (2010b), EBB (2009), own calculations.

Figure 22. Development of plant oil demand for biodiesel production and provision of plant oil in Germany (rape seed production and import converted to equivalent amount of plant oil).

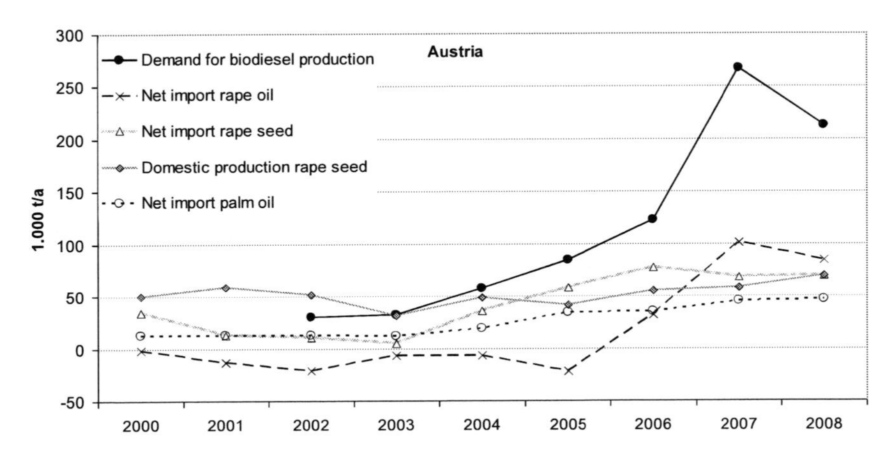

Source: UN Comtrade (2009), Eurostat (2010b), EBB (2009), own calculations.

Figure 23. Development of plant oil demand for biodiesel production and provision of plant oil in Austria (rape seed production and import converted to equivalent amount of plant oil).

This supports the presumption by Rosillo-Calle et al. (2009), who argue that "increasing consumption of domestically produced rapeseed oil for biodiesel uses may have led to a considerable gap in EU food oil demand (which continues to increase), resulting in an increase on imports for other types of oil (mostly edible palm oil)".

We conclude that the increasing biodiesel production in Germany and Austria led to significant shifts in international trade of plant oil and oil seeds. As shown in section 2.2.2, progress in the field of biofuels for transport was very uneven among CE countries. Therefore, the additional crop demand could initially be imported from neighbouring countries with favourable conditions for increasing energy crop production, especially Czech Republic and Hungary.

However, recent data suggest that with the demand for energy crops also increasing in these countries, imports from other European, or especially Non-European countries are getting inevitable. It was already shown in section 3.2.4 that the CE region recently turned from a net exporter of biofuels into a net importer. With regard to the supply of oilseeds, the data are even more striking: The total net imports of rape seed to CE increased from 0.76 Mt in 1996 to about 2 Mt in 2008.[22]

[22]Slovakia was excluded from this calculation due to highly implausible data for 2008 (UN Comtrade, 2009).

4. Potentials and Prospects for an Enhanced Use of Bioenergy in Central Europe

4.1. Bioenergy in the Context of Eu Energy Policy

With the implementation of the 2009-RES-Directive (EC, 2009a) an "overall binding target of a 20% share of renewable energy sources in energy consumption [...] as well as binding national targets by 2020 in line with the overall EU target of 20%" have been established. The share of RES is calculated as the sum of final energy from RES consumed in the heat, transport and electricity sector, divided by the total final energy consumption, including distribution losses and consumption of the energy sector. In addition to the overall 20% target by 2020, a sub-target for the transport sector (including road and rail transport) in the amount of 10% was defined. Renewable electricity used in electric cars is also taken into account; in consideration of the higher efficiency of electric drivetrains, a factor of 2.5 is applied for electric cars. In order to promote advanced biofuels produced from nonfood cellulosic materials and ligno-cellulosic materials, the amounts of "advanced" biofuels count twice towards the target. Still, the main contribution towards the sub-target is expected to come from biodiesel and ethanol.

In the European Biomass Action Plan (EC, 2005) it is recognized that bioenergy is of major importance for increasing the share of renewable energies and reducing dependence on energy imports. The projections made for the Renewable Energy Road Map (EC, 2006) suggest that the use of biomass can be expected to double and to contribute around half of the total effort for reaching the 20% target.

The "strengthened national policy scenario" in Resch et al. (2008a) gives an impression of to what extent bioenergy can contribute towards fulfilling the 2020-targets in CE.[23] The scenario is based on the following core assumptions: The implementation of "feasible" energy efficiency measures (leading to a moderate development of the future overall energy demand as projected in the PRIMES target case (Capros et al., 2008). Support conditions for RES are improved, leading to the fulfillment of the EU-wide 20%-target by 2020.

[23]The scenarios have been compiled with the simulation tool Green-X. This model simulates future investments in renewable energy technologies for heat, electricity and transport fuel production, based on a myopic economic optimization. The availability of biomass resources, cost and price developments, the energy demand and its structure, diffusion and other influencing parameters as well as energy policy instruments are considered within the simulation runs.

This simulation confirms that biomass is of crucial importance for meeting the 2020-targets. In all CE countries more than 50% of the growth in RES until 2020 is made up by bioenergy.

In the Czech Republic, Hungary, Poland and Slovakia bioenergy even accounts for more than 75% of the growth. The consumption of biomass as share of the total GIC according to this scenario range from 7.6% in Italy to 25.3% in Denmark.

Table 2. shows a summary of the share of biomass and all RES in the total energy consumption in the reference year 2005 and 2007 (the latest year available in statistics), the national targets according to the 2009-RES-Directive and the contribution of biomass according to the "strengthened national policy scenario" in Resch et al. (2008a).

Table 2. Summary of the current state, targets and prospects for the share of biomass and RES in CE countries (all values in %)

Concept	Reference	Fraction, year	AT	CZ	DE	DK	HU	IT	PL	SI	SK
Final energy consumption	EC (2009a)	RES, 2005	23.3	6.1	5.8	17.0	4.3	5.2	7.2	16.0	6.7
		RES Target, 2020	34.0	13.0	18.0	30.0	13.0	17.0	15.0	25.0	14.0
Gross inland consumption	Eurostat (2010a), own calculations	RES, 2005	21.7	4.0	5.1	16.4	4.4	6.5	4.8	10.6	4.3
		RES, 2007	23.8	4.7	8.3	17.4	5.3	6.9	5.1	10.0	5.5
		Biomass [a], 2005	11.1	3.5	3.4	12.4	3.9	1.9	4.6	6.5	2.1
		Biomass [a], 2007	13.3	4.2	5.7	13.2	4.6	2.2	4.8	6.2	3.2
	Resch et al. (2008a)	Biomass scenario 2020	22.0	9.0	10.2	25.3	11.2	7.6	13.5	15.1	9.6

Notes: a)Non-renewable wastes have been deducted based on Eur'ObservER (2010).

4.2. Review and Discussion of Biomass Potentials in Literature

Assessments of biomass supply potentials are numerous and the results vary widely. There are different concepts of potentials like "theoretical", "technical" or "environmentally compatible" potentials (see Rettenmaier et al., 2008). Potentials in literature are usually qualified according to these definitions. Yet methodological approaches, assumptions and constraints of potential assessments differ from study to study.

The following analyses are based on three studies (EEA, 2006; Thrän et al., 2005 and de Wit and Faaij, 2010) which have been chosen for the following reasons: Uniform methodologies have been applied, they comprise all types of biomass resources (with the exception of non-agricultural residues not being considered in de Wit and Faaij, 2010) and results are available for all CE countries, broken down by country and biomass type. The main features of the methodological approaches applied and databases used are summarized in Table 3.

According to the Eurostat definition of "biomass consumption", biofuels for transport are represented with the calorific value of the fuel (and not with the amount of biomass used to produce the fuel). Due to the relatively low conversion efficiencies (e.g. typically 55% for ethanol and 57% for biodiesel[24]; cp. AEBIOM, 2007) the energy content of the quantity of feedstock used for the production (primarily energy crops) is clearly higher than the consumption according to energy statistics (and shown in Figure 4). This needs to be taken into account when comparing statistical data with data on biomass supply potentials.[25]

The methodological approaches of the considered studies are basically quite similar. The most significant differences include environmental restrictions considered, scenario assumptions and influencing factors which are taken into account as well as assumptions about energetically usable fractions of certain biomass resources. Figure 24 shows a comparison of the results. The biomass production and consumption in the year 2007 are also included.

[24]The conversion efficiencies stated here are defined as the ratio of the energy content of the biofuel to the primary energy content of the feedstock used, with by-products (which can be used for energy recovery, for feed or material uses) not taken into account.

[25]That biofuels for transport are represented with the calorific value of the fuel and not with the primary energy required for the production of the biofuel is still justified by the following facts: First, this allows for a direct comparison of fossil fuel and biofuel consumption and second, the above mentioned by-products are thereby rightly excluded from the statistics.

Table 3. Summary of features, references/databases used and methodologies applied for assessing biomass potentials in Thrän et al. (2005), EEA (2006) and De Wit and Faaij (2010)

	Thrän et al. (2005)	EEA (2006)	De Wit and Faaij (2010)
Type of potential	Technical potential with consideration of structural and ecological restrictions	Technical potential with consideration of environmental criteria ("Environmentally-compatible potential")	"Supply potential" (forest biomass: sustainable potential)
Reference years	2000, 2010, 2020	2010, 2020, 2030	2030
Methodological approaches and main references/databases			
Methodology for assessing forest biomass potential	Comprises potential from current use (felling residues) and potential from annual increment (annual growth minus fellings) Base year: 2000 2010 and 2020: Increasing demand for wood products according to UNECE (2000) Main databases: FAO (2010c), FAO (2005), UNECE (2000)	Comprises "residues from harvest operations normally left in the forsest ("felling residues") and complementary fellings" Complementary fellings describe difference between maximum sustainable harvest level and actual harvest needed to satisfy roundwood demand Environmental considerations include biodiversity, site fertility, soil erosion, water protection Criteria to avoid increased environmental pressure applied Databases: FAO (2010c), OECD Europe (projections for wood demand)	Comprises "difference between actual felling and felling residues and the net annual increment" (including stems) Main database: Karjalainen (2005)
Methodology for assessing potential of biogenic wastes and residues	Comprises only residues which are not usable for material uses Exemplary proportions assumed to be available for energy recovery: sawmill residues 10%, bark 80%, waste wood 75% (estimated on basis of per capita production), straw 20% of total production Other potentials based on scenarios and assessments in literature as well as estimates: e.g. manure based on livestock scenarios and assumptions about husbandry conditions, black liquor based on rough assessments and other studies, food processing industries also considered Databases: FAO (2010c), Eurostat (2010b)	Comprises solid and other agricultural residues, manure, biogenic fraction of municipal solid waste (MSW), black liquor, wood-processing waste wood, construction and demolition wood, other waste wood, sewage sludge and food processing wastes Environmental criteria assumed: waste minimization, no energy recovery from waste currently going to recycling or reuse (estimated proportions), production of timber and wood products declines, extensive farming practices etc. Projections for waste fractions based on different scenarios in literature (e.g. FAO, 2005; Skovgaard et al., 2005)	Comprises only agricultural residues obtained during production of food and feed Crop-specific ratio of crop residue to crop main produce applied Assumed "residue use factor": 50% Main database: FAO (2010c)

Table 2. (Continued)

	Thrän et al. (2005)	EEA (2006)	De Wit and Faaij (2010)
Methodology for assessing potential of dedicated energy crops	Base year: 2000 (average over 3 to 5 years) Evaluation of surplus arable land and grassland available for dedicated energy crop production Reduction of production surplus and related exports assumed Considered influencing factors: population scenarios, reduction of agricultural land, yield increases, increasing efficiency in livestock breeding Assumed distribution of energy crops Databases: FAO (2010c), Eurostat (2010b)	Evaluation of released and set-aside land under assumption of further reform of common agricultural policy (based on EuroCare, 2004) Competition effect between bioenergy and food production are only taken into account for Germany Assumption of site-specific environmentally-compatible crop mixes Increase in crop yields according to EuroCare (2004) Environmental criteria assumed: 30% of agricultural land dedicated to environmentally-oriented farming, 3% set aside land, extensively cultivated agricultural areas are maintained, bioenergy crops with low environmental pressure are used	Evaluation of surplus arable land and grassland available for dedicated energy crops Projected changes in land area requirements (population size, dietary habit, agricultural productivity, self-sufficiency ration of Europe) Assumption: Europe maintains current food and feed self-sufficiency of about 90% Different assumptions for yield increases and different sustainability criteria assumed Databases: Fischer et al. (2010b)

Basically it can be concluded that there are substantial unused biomass potentials in all CE countries. While forest biomass and biogenic wastes remain fairly constant, the potential of dedicated energy crop production is assumed to become increasingly important.

The potentials of biogenic wastes are the most consistent throughout the studies. This is unsurprising since they are based on current production statistics and often the same databases were used. However, it should be considered that the potentials of waste and residues are essentially based on estimated “use factors”.

In the case of straw, this use factor is assumed 20% in EEA (2006) and 50% in de Wit and Faaij (2010). As the latter point out, the amount of straw which can be removed and used energetically without causing adverse environmental effects is actually site-specific and depends on numerous factors. Highly aggregated assessments of biogenic wastes can therefore only be seen as rough estimates. In order to derive profound data, detailed bottom-up approaches are required, carried out in the course of regional energy concepts, for example. Another aspect to be considered in connection with the assessment of waste and residue potentials based on production statistics is that they sometimes include significant amounts of indirectly imported biomass. For example in Austria the potential of wood processing residues is to a large extent based on imported roundwood (see section 3.4.3). Strictly speaking, this fraction cannot be considered a domestic biomass potential.

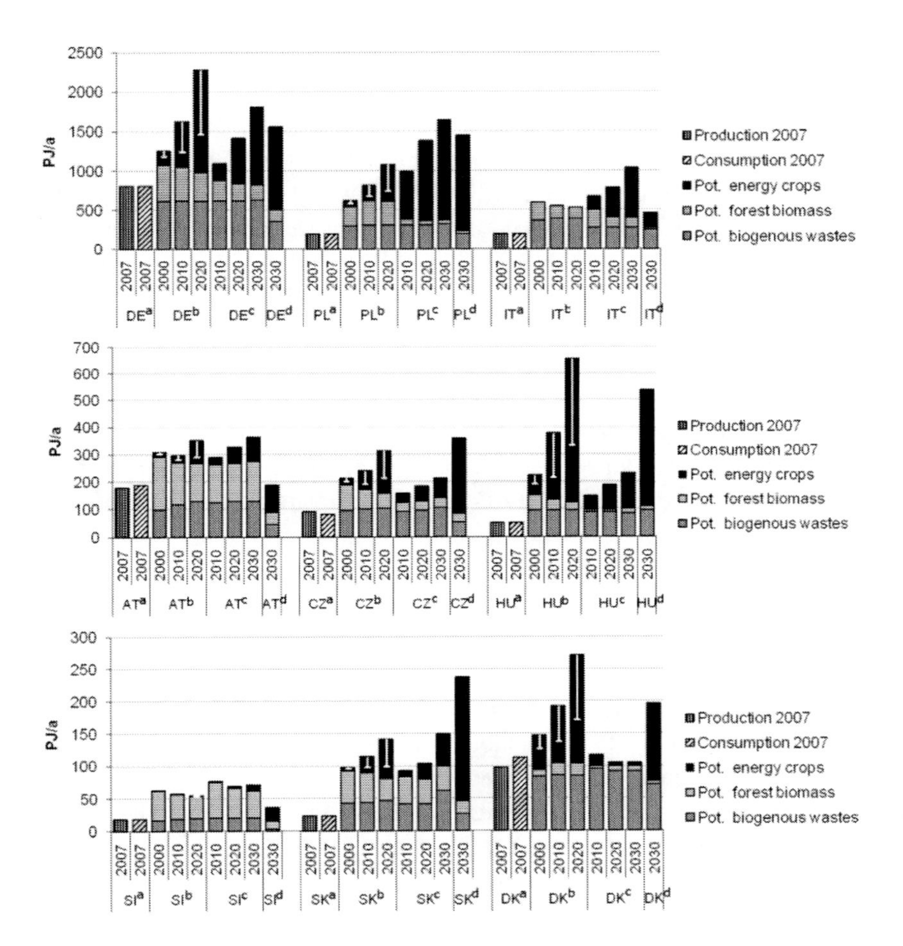

Source: a) Eurostat (2010a), b) Data obtained from Thrän et al. (2005) (error bars represent results for the environmentally-oriented scenario), c) Data obtained from EEA (2006), d) Data obtained from de Wit and Faaij (2010) (baseline scenario; biogenic wastes comprise only agricultural residues).

Figure 24. Comparison of biomass production and consumption in 2007 with biomass potentials ("Pot.") according to three studies.

The increasing potentials of energy crops are on the one hand due to assumed yield increases in both energy crop and food and feed production, and on the other due to scenario assumptions for the future development of agricultural production in Europe. Model-based simulations of the agricultural developments in the EU (e.g. of the CAPSIM model used in EEA, 2006) indicate that with continuing reforms of the common agricultural policy resulting in gradual liberalization of agricultural markets and a reduction in subsidized exports, agricultural productivity can be increased significantly and the current self-sufficiency for food and feed products maintained with clearly less agricultural land. Thus, surplus land is assumed to be made available for energy crop production.

To what extent the consideration of different environmental criteria influence the supply potentials of energy crops is illustrated by the environmentally-oriented scenario according to Thrän et al. (2005), represented by the error bars in Figure 24. The Low and High estimate scenarios in de Wit and Faaij (2010) illustrate that assumptions about yield increases have a huge impact on energy crop potentials. Furthermore, especially with regard to the energy crop potentials in Poland, Italy, Hungary and Denmark there are also significant inconsistencies which cannot be explained easily, indicating that there are substantial uncertainties connected with the future potential of energy crops. The potential of forest biomass primarily depends on the currently unused annual growth. Furthermore, scenarios for the demand of wood products and the development of the wood-processing industries have a major impact. A comparison between EEA (2006) and Thrän et al. (2005) indicates that the additional environmental criteria considered in the former result in a significant reduction of the forest biomass potential. Regardless of the uncertainties related to potential assessments, the following conclusions are drawn: Only in Germany, Austria and Denmark more than half of the biomass supply potential was actually utilized in 2007. The structure of biomass potentials is highly inhomogeneous. According to these studies, especially Germany, Poland and Hungary are capable of increasing the energy crop production substantially, while maintaining the current self-sufficiency for food and feed. The potential of forest biomass is generally rather limited, partly due to the increasing wood demand of the wood-processing industries. Biogenic wastes and residues, including waste wood, wood processing and agricultural residues as well as residues from food processing are a considerable potential. The figures indicate that in several CE countries, the potential of wastes and residues even surpass the total biomass production in 2007.

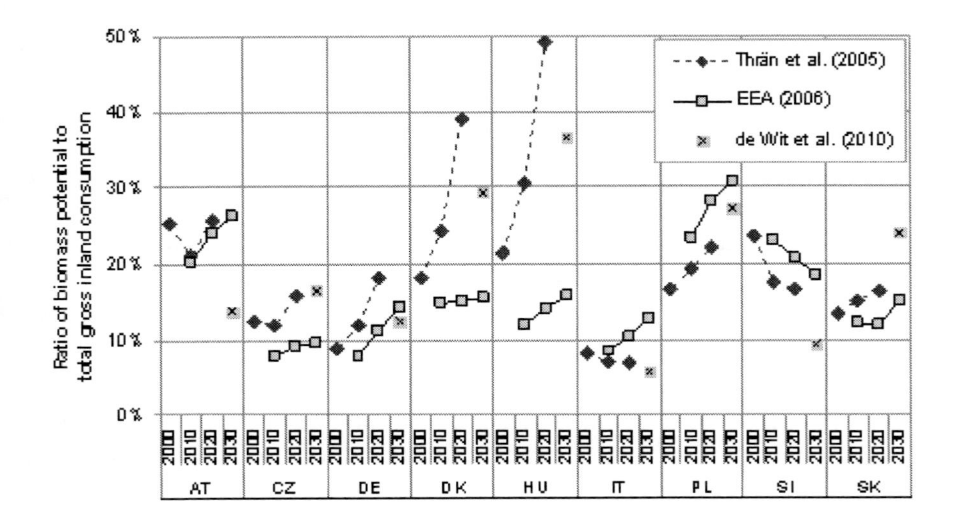

Figure 25. Ratio of biomass supply potential according to Thrän et al. (2005), EEA (2006) and de Wit and Faaij (2010) to total gross inland energy consumption (scenario according to Capros et al., 2008; "PRIMES target case").

Figure 25. shows the biomass potentials of the considered studies as shares of the GIC (projections according to Capros et al., 2008). A comparison with Table 2 (scenarios according to Resch et al., 2008a) reveals that Poland could act as the main exporter of biomass in CE. Even if Poland's 2020-target is primarily achieved with biomass (as projected in Resch et al., 2008a) the unused biomass potential accounts for approximately 500 PJ. In most other countries the domestic biomass potential needs to be utilized to a large extent to fulfil the 2020-targets. With regard to Germany, Italy, Denmark and Hungary no definite conclusions can be drawn due to big uncertainties as to what extent the supply potential of biomass can be extended with the production of energy crops.

5. Discussion, Conclusion and Policy Implications

5.1. Recent Developments in Bioenergy Use in CE

Bioenergy is currently the most important source of renewable energy in CE. The contribution of biomass and wastes to the total energy supply (gross inland consumption) in CE countries ranges from 2.8% in Italy to 14.9% in Denmark (2008).

European directives and according national support schemes have already led to significant progress in recent years. Progress was very uneven in the considered countries. The CE countries with the highest growth of biomass as share of the GIC from 2000 to 2008 were Denmark (+6%), Germany (+4.8%), Austria (+4.5%) and Hungary (+3.9%). It is remarkable that the countries which already had the highest bioenergy shares in 2000, namely Austria and Denmark, are among these countries. It is therefore concluded that at least in recent years, the crucial barriers for an increase in bioenergy use was not the availability of biomass resources in the CE region but the typical barriers for upcoming technologies, such as know-how, capacity building of equipment etc. In absolute numbers, Germany showed by far the highest increase in bioenergy use. In 2008 Germany was accountable for more than 50% of the total biomass consumption and production in the considered countries, and therefore dominates the structure of the energetic biomass use in CE.

Even though heat generation is the oldest and often most competitive utilization path for biomass, EU Directives as well as national support schemes were focused on the electricity and transport sector in recent years. As a result, the annual increments in biomass-based heat generation have been relatively stable since 1990, whereas in the field of power and CHP generation and the production of transport fuels, growth rates increased considerably after the year 2000. It is assumed that as a consequence of the 2009-RES-Directive (EC, 2009a), in which national targets for the share of RES in the final energy consumption are defined, more attention will be paid to biomass use in the heat sector in the years to come.

5.2. International Biomass Trade

The challenges related to mapping international trade streams of biomass for energy are numerous, and assessing the impact of the growing bioenergy use on trade streams is not straightforward. To this end, specific methodologies need to be developed, especially when it comes to assessing indirect effects like spillover effects or indirect land-use change.

Based on the approaches applied in this work, it is concluded that the main importers of wood fuels in CE are Italy, Denmark and Austria. Cross-border trade of wood pellets has increased significantly in recent years and is already of high importance for the Danish bioenergy sector. (Pellets represent by far the most important fraction of biomass imports to Denmark.) Austria, being a net exporter of wood pellets, is importing considerable amounts of wood residues, primarily indirectly in the form of industrial roundwood.

The comprehensive assessment of biomass trade related to bioenergy carried out for the case of Austria indicates that indirect net-imports of wood-based fuels are more significant than direct trade, and that feedstock imports for biofuel production are roughly as important as direct biofuel trade. Hence, it is clearly insufficient to rely only on energy statistics (which do not include indirect trade streams and cross-border trade of feedstock used for biofuel production) when assessing international trade related to bioenergy use.

With regard to direct biofuel trade, Austria, Italy and Poland are the main importers (primarily biodiesel). Although growing rapidly, cross-border trade related to biofuels for transport is still rather moderate compared to (indirect and direct) trade of wood fuels in CE. Still, as more and more (Central) European countries aim at achieving their biofuel targets, it is either necessary to mobilize domestic resource potentials or further increase imports from Non-European countries. There is strong evidence that the CE region is currently becoming increasingly dependent on imports of biofuels as well as feedstock for biofuel production. There is also evidence that in Germany and Austria (which are most advanced in biofuel use), the growing demand for plant oil for biodiesel production primarily resulted in an increase in imports rather than the mobilization of domestic potentials (also palm oil imports have been increasing, albeit to a rather limited extent). Thus, in order to avoid adverse effects of the enhanced use of biomass (especially indirect land-use change), the need for obligatory certification schemes for sustainably produced biomass is becoming increasingly urgent. The enhancement of international biomass trade often seen as a key factor for mobilizing the (global) biomass supply potential, avoiding short-term regional supply problems and providing the framework conditions required for steady growth of bioenergy use. However, concerns about sustainability issues of globally traded biomass resources have to be taken seriously, and in order to enhance the security of supply and facilitate domestic income, a main focus of national biomass action plans should be put on the mobilization and use of regional biomass resources.

5.3. Resource Potentials

It is apparent that there are numerous aspects and barriers for an enhanced use of biomass, which cannot be considered in highly aggregated assessments of biomass potentials. Therefore, the assessment of locally available residues and wastes as well as specific measures for their utilization should be promoted in regional energy concepts and action plans. Increasing biomass imports to countries with a rapid growth of the bioenergy sector on the one hand, and evidence of unused domestic resource potentials on the other indicate that the supply with regional biomass has not been given enough attention within energy policy strategies, according support schemes and incentives. In particular, it should be investigated whether the cost of regional supply chains can be decreased with logistical improvements, the enhanced use of conversion technologies (e.g. pelletizing, torrefaction) and removal of organisational barriers. Results of studies on biomass resource potentials indicate that there are vast unused potentials in most CE countries. According to EEA (2006) the environmentally compatible potential in the year 2010 in the considered countries is about two times higher than the current utilization (2007), and the potential in 2030 even three times higher. The results of other studies show even higher supply potentials. The consideration of different environmental criteria has a significant impact on the amount of agricultural and forest biomass potentials, indicating that there is a considerable risk that uncoordinated growth of bioenergy use results in additional pressure on the environment. The consideration of environmental criteria in the design of bioenergy support schemes (especially promoting the mobilization of biomass resources) is therefore of crucial importance. To what extent the biomass potentials are already utilized is highly diverse among CE countries: In Denmark, Germany and Austria the currently unused resource potential is relatively small, whereas countries like Poland, Italy and Slovakia only use a very low proportion of their biomass potential. Especially agricultural resources (including energy crops as well as residues and wastes) are assumed to constitute a substantial potential that is hardly tapped yet. It is assumed that to some extent, the very uneven progress in biomass use (primarily resulting from diverging energy policies, support schemes and as a consequence diverging biomass price developments) encouraged cross-border trade between European countries. Increasing efforts in the field of bioenergy throughout all EU countries are likely to result in a further shift of trade flows towards international (trans-continental) biomass trade.

5.4. Towards the 2020-Targets

The importance of bioenergy for reaching the 2020-targets defined to the 2009-RES-Directive is undisputed. Scenarios by Resch et al. (2008a) indicate that among the renewable sources of energy, biomass can be expected to bring the biggest contribution to the achievement of the 2020-targets. Special attention should therefore be attributed to the design of support schemes promoting bioenergy use. Aspects which should be considered within national biomass action plans include the following: Biomass can be used in all energy sectors (heat, electricity and transport) and the economic and environmental properties of the different bioenergy utilization paths often vary widely. Clear strategies and targets for the development of the bioenergy sector, designed with consideration of technological, economic and ecological criteria are essential (see Kalt et al., 2010).

Finally, it has to be taken into account that increasing competition for biomass resources between the different types of biomass use (both for energy and material uses) are expected with the progressing exploitation of biomass potentials. In order to facilitate the diffusion of the most efficient utilization paths, bioenergy policies should be designed to counteract resource competition as far as possible; both with supply-side measures and clear priorities for the most beneficial technologies and utilization paths.

REFERENCES

AEBIOM, 2007. European Biomass Association, European Biomass Statistics 2007, A statistical report on the contribution of biomass to the energy system in the EU 27, Brussels.

AGRANA, 2010. Financial report (Jahresfinanzbericht) 2009/10, AGRANA Beteiligungs-AG, Vienna.

Austrian Chamber of Agriculture, 2010. Holzmarktbericht (Wood market report) 2005 – 2009, Vienna.

Austropapier, 2010. Website of the Association of the Austrian Paper Industry, Statistics - raw materials, http://www.austropapier.at/index. php?id= 81andL=1, last access in December 2010

Bloomberg New Energy Finance, 2010. Carbon Markets – North America – Research Note. *A fresh look at the costs of reducing US carbon emissions*; 14 January 2010.

BMLFUW, 2009. Austrian Federal Ministry of Agriculture, Forestry, Environment and Water Management, Biokraftstoffe aktuell - Zahlen und Fakten (Facts and figures on biofuels), http://umwelt.lebensministerium. at/article/articleview/66083/1/1467, last access in January 2011.

BMLFUW, 2010. Austrian Federal Ministry of Agriculture, Forestry, Environment and Water Management, Holzeinschlag (Austrian timber felling report) 2009, Vienna.

Capros P., Mantzos L., Papandreou V., Tasios N., 2008. European Energy and Transport – Trends to 2030, Update 2007, European Commission, Directorate-General for Energy and Transport, Institute of Communication and Computer Systems of the National Technical University of Athens.

De Wit M.P., Faaij A.P.C., 2010. *European Biomass Resource Potential and Costs, Biomass and Bioenergy* 34 (2010) 188 – 202.

Danish Energy Agency (DEA), 2009. *Danish annual report under the Biofuels Directive* (Directive 2003/30/EC).

EBB, 2011. Website of the European Biodiesel Board, Statistics, http://www.ebb-eu.org/stats.php, last access in January 2011.

ePURE, 2011. Website of the European Bioethanol Fuel Association, Statistics, updated September 2009, http://www.epure.org/theindustry/ statistics, last access in January 2011.

EEA, 2006. T. Wiesenthal, A. Mourelatou, J.-E. Peterson, European Environment Agency, P. Taylor, AEA Technology, How much bioenergy can Europe produce without harming the environment?, EEA Report No 7/2006, Copenhagen.

ENS, 2009. Danish Energy Agency, Energy Statistics 2008, Copenhagen.

EurObserv'ER, 2009. The State of Renewable Energies in Europe, 9th EurObserv'ER Report, Observ'ER (F), Eclareon (DE), "Jožef Stefan" Institute (SI), Energy research Centre of the Netherlands (NL), Institute for Renewable Enegy (IEO/EC BREC, PL) http://www.eurobserv-er.org/, last access in January 2011.

European Commission (EC), 2001. Directive on electricity production from renewable energy sources, COM 2001/77/EC.

European Commission (EC), 2003. Directive on the promotion of the use of biofuels or other renewable fuels for transport, COM 2003/30/EC.

European Commission (EC), 2005. Biomass Action Plan, COM(2005) 628.

European Commission (EC), 2006. Renewable Energy Road Map, Renewable energies in the 21st century: building a more sustainable future, COM(2006) 848.

European Commission (EC), 2007. Commission Regulation (EC) No 1214/2007 amending Annex I to Council Regulation (EEC) No 2658/87 on the tariff and statistical nomenclature and on the Common Customs Tariff.

European Commission (EC), 2009a. Directive 2009/28/EC of the European parliament and of the council on the promotion of the use of energy from renewable sources and amending and subsequently repealing Directives 2001/77/EC and 2003/30/EC.

European Commission (EC), 2009b. The Renewable Energy Progress Report, Commission Report in accordance with Article 3 of Directive 2001/77/EC, Article 4(2) of Directive 2003/30/EC and on the implementation of the EU Biomass Action Plan COM(2005) 628, COM(2009) 192.

European Commission, 2010a. National reports on the implementation of Directive 2003/30/EC of 8 May 2003 on the promotion of the use of biofuels or other renewable fuels for transport, http://ec.europa.eu/energy/renewables/biofuels/ms_reports_dir_2003_30_en.htm, last access in December 2010.

European Commission, 2010b. Website of the European Commission, Climate Action, Policies, Emissions Trading System, http://ec.europa.eu/clima/policies/ets/index_en.htm, last access in December 2010.

European Commission, 2010c. Report from the Commission on indirect land-use change related to biofuels and bioliquids, COM(2010) 811, Brussels.

Eurostat, 2010a. Website of Eurostat, energy statistics, http://epp.eurostat.ec.europa.eu/portal/page/portal/energy/data/database, last access in December 2010.

FAO, 2005. European Forest Sector Outlook Study: 1960-2000-2020, Main Report by the Food and Agriculture Organisation (FAO) of the United Nations, Geneva Timber and Forest Study Paper 20.

FAO. 2010a. Statistical database of the Food and Agricultural Organisation (FAO) of the United Nations, ForesSTAT Database, http://faostat.fao.org/site/626/default.aspx#ancor, last access in December 2010.

FAO, 2010b. Statistical database of the Food and Agricultural Organisation (FAO) of the United Nations, Forestry Trade Flows, http://faostat.fao.org /site/628/default.aspx, last access in December 2010.

FAO, 2010c. Statistical database of the Food and Agriculture Organization (FAO) of the United Nations, FAOSTAT Database, http://faostat.fao.org /default.aspx, last access in December 2010.

Fischer G., Hizsnyik E., Prieler S., Shah M., van Velthuizen H., 2010a. Biofuels and Food Security, OPEC Fund for International Development (OFID) study prepared by International Institute for Applied Systems Analysis (IIASA), Laxenburg.

Fischer G., Prieler S., van Velthuizen H., Berndes G., Faaij A.C.P., Londo M., de Wit M., 2010b, Biofuel production potentials in Europe, sustainable use of cultivated land and pastures. Part II: land use scenarios, Biomass and Bioenergy 34 (2010), 173–187.

Freshfields Bruckhaus Deringer, 2008. Briefing: Import of biodiesel into the European Union, Customs law classification of B100 and B99, January 2008.

Hagauer D., Lang B., Nemestothy K., 2007. Woodflow Austria 2005, Austrian Energy Agency, BMLFUW, Klima:aktiv Energieholz, http://www.klimaaktiv.at/filemanager/download/30224, last access in December 2010.

Heinimö J., Pakarinen V., Ojanen V., Kässi T., 2007. International bioenergy trade - scenario study on international biomass market in 2020, Lappeenranta University of Technology, Lappeenranta.

Heinimö J., Junginger M., 2009. Production and trading of biomass for energy – An overview of the global status, Biomass and Bioenergy 33 (2009) 1310-1320.

IEA, 2010a. Website of IEA Task 40: International bioenergy trade http://www.bioenergytrade.org/, last access in March 2010.

IEA, 2010b. Website of IEA Bioenergy, http://www.ieabioenergy.com/, last access in March 2010

Kalt G., Kranzl L., Haas R., 2010. *Long-term strategies for an efficient use of domestic biomass resources in Austria, Biomass and Bioenergy* 34 (2010) 449 – 466.

Karjalainen T., Asikainen A., Ilavsky J., Zamboni R., Hotari K.E., Röser D., 2005. *Estimation of energy wood potential in Europe*. Finnish Forest Research Institute.

Kopetz H., Moidl S., Prechtl M., Kirchmeyr F., Kronberger H., Kanduth R., Rakos C., 2010. Nationaler Aktionsplan für erneuerbare Energie (National action plan for renewable energies), Technical report, Vienna.

Peksa-Blanchard M., Dolzan P., Grassi A., Heinimö J., Ranta T., Junginger M., Walter A., 2007. Global Wood Pellets Markets and Industry: Policy Drivers, Market Status and Raw Material Potential, IEA Bioenergy Task 40.

Pellet@las, 2010. Website of the Pellet@las project, http://www.pelletcentre.info/cms/site.aspx?p=9107, last access in February 2010.

Resch G, Faber T, Panzer C, Haas R, Ragwitz M, Held A, 2008a. Futures-E, 20% RES by 2020 – A balanced scenario to meet Europe's renewable energy target, Intelligent Energy for Europe-Programme, Vienna.

Resch G., Panzer C., Coenraads R., Reece G., Kleßmann C., Ragwitz M., Held A., Konstantinaviciute I., Chadim T., 2008b. Renewable Energy Country Profiles, TREN/D1/42-2005/S07.56988, Utrecht.

Rettenmaier N., Reinhardt G., Schorb A., Köppen S., von Falkenstein E. et al., 2008. Status of biomass resource assessments – Version 1, Biomass Energy Europe project, IFEU, Heidelberg.

Rosillo-Calle F., Pelkmans L., Walter A., 2009. A global overview of vegetable oils, with reference to biodiesel, IEA Bioenergy Task 40.

Schmied A., 2009. personal information by DI Alexander Schmied on the wood consumption of the Austrian wood board industry, Association of the Austrian Wood Industries, Oktober 2009, Vienna.

Skovgaard M., Moll S., Møller Andersen F., Larsen H., 2005. Outlook for waste and material flows: Baseline and alternative scenarios, ETC/RWM working paper 2005/1, EEA.

Statistik Austria, 2010a. Energy balances 1970 bis 2008 (detailed information), Vienna. http://www.statistik.at/web_en/statistics/energy_ environment/energy /energy_balances/index.html, last access in December 2010.

Statistik Austria, 2010b. Overall energy consumption of households, http://www.statistik.at/web_en/statistics/energy_environment/energy/energy_ consumption_of_households/index.html#index1, last access December 2010.

Statistik Austria, 2010c. Supply balance sheets for the crop sector, http://www.statistik.at/web_en/statistics/agriculture_and_forestry/prices_bala nces/supply_balance_sheets/index.html, last access in December 2010.

Thrän D., Weber M., Scheuermann A., Fröhlich N., Zeddies J., Henze A., Thoroe C., Schweinle J., Fritsche U., Jenseit W., Rausch L., Schmidt K., 2005. Nachhaltige Biomassenutzungsstrategien im europäischen Kontext, Leipzig.

UN Comtrade, 2009. United Nations Commodity Trade Statistics Database (UN Comtrade database) http://comtrade.un.org/db/, last access in June 2009.

UNECE, 2000. United Nations Economic Commission for Europe, Food and Agricultural Organisation of the United Nations, *Temperate and Boreal Forest Resource Assessment* (TBFRA-2000).

In: Energy Resources
Editor: Enner Herenio de Alcantara

ISBN: 978-1-61324-520-0

Chapter 3

EMERGENCY PLANNING ZONE: CONSTRAINTS AND OPPORTUNITIES FOR THE DEVELOPMENT OF NUCLEAR ENERGY AND EXPLOITATION OF ITS PROCESS HEAT

Giorgio Locatelli[*] and Mauro Mancini
Politecnico di milano department of Management,
Economics and Industrial Engineering Via Lambruschini, Milano, Italy

ABSTRACT

Light Water Reactors (LWR), which represent the most common reactor in operation and under construction, have an average thermal efficiency of about 33%-35%, therefore two third of the thermal energy produced by the nuclear reaction is typically wasted. The literature presents many possible applications of this thermal energy, however most of them are not feasible because of economic and legislative constraints: among the others the Emergency Planning Zone (EPZ) is one of the most critic. The EPZ is the area surrounding the Nuclear Power Plants (NPP) subject to specific rules constraining the development of the area. These constraints avoid the complete exploitation of the energy produced by the power plants.

Small Medium Nuclear Reactors (SMR) can offset some of the constraints since they are intrinsically safer and therefore can theoretically require a smaller EPZ. This chapter deals with the relationships among the

[*]Tel: +39 02 2399 4096 Fax: +39 02 2399 4083 E-mail: giorgio.locatelli@polimi.it

Emergency Planning Zone (EPZ), the reactor size and the possibilities of cogeneration. After a review of the constraints for the EPZ in the different countries it show the relationship among EPZ and NPP size and presents the various options of nuclear cogenerations. These options are evaluated according to the commercial feasibility of the different technologies clustering the solutions in short and long term options.

1. Introduction

A nuclear power plant (NPP) considered as a whole presents some by-products that can be exploited to create interesting synergies between the plant itself and potential nearby facilities. These by-products can be ascribed directly or indirectly to the nuclear plant:

- Directly if they derive from the nuclear reactor itself;
- Indirectly if they are due to the presence of the nuclear power station and, consequently, to the features of the location, the need of ancillary installations or the safety measures required by the NPP.

The main issue is whether - and how - these by-products can be harnessed in order to increase both the economic attractiveness and the social acceptability of the nuclear power plant. The average electric efficiency of a Light Whater Reactor (LWR), the most common technology for existing reactor and proposed plant (as EPR, AP1000, ABWR, IRIS etc...) is about 33%. These two thirds of the thermal power produced by the reactor are wasted in the environment while converting heat into electricity. The "wasted" heat can be used in a co-generation mode for several purposes, depending mainly on the outlet temperature of the reactor (and, consequently, on the reactor type). Co-generation allows achieving overall efficiencies (thermal and electrical) up to 85%. This means that the primary energy used to produce at the same time heat and electricity is much lower than the primary energy that would be required to produce separately the same amounts of heat and electricity. Moreover, the cost of heat production is lower if compared to the separate mode, because of the availability of an almost free heat source. Finally, the consumption of fossil fuels for heat production is strongly reduced: this leads to reduced greenhouse gas emissions. Depending on the reactor type, it is possible to combine applications operating at low or high temperature. Low temperature applications basically belong to three categories: district heating (for residential, commercial or agricultural use), desalination and process heat delivery to factories with low-temperature requirements. High temperature applications are more innovative and can be subdivided into: hydrogen production (in order to store and distribute energy), process heat for industries that operate at high temperatures, oil shale extraction and biomass gasification or other fuel syntheses. Potential nuclear heat applications are described in section □. Nuclear legislation prescribes mandatory safety measures, such as Emergency Planning Zones (EPZs) and site selection criteria. As a consequence, NPPs are usually placed in scarcely populated areas, surrounded by kilometres of unused land. In particular, the EPZ around a nuclear plant causes major impediments to human activities and industrial uses of the site, as it imposes measures such as the possibility of a total evacuation in case of an accident. The current challenge is to reduce the width of emergency zones, making them proportional to the safety and size of the reactor. Therefore, EPZs for new reactors belonging to Generations III+ and IV are likely to have a smaller extension than traditional reactors, however, this is not sure. Moreover, nuclear legislation differs from country to country, and some governments could decide to maintain standard EPZ sizes even for innovative reactors. If such reductions won't be possible, the mentioned areas should then be exploited in the most profitable way. These areas can be employed with applications that require wide extensions of ground but that do not require a high human density, such as industrial parks, or farming installations for the cultivation of energy crops, or renewable energy generation plants, such as photovoltaic or wind turbines, depending on the meteorological conditions of the site. Particular attention has been posed on the EPZ topic, which is developed in the next sections. In order to identify possible applications to harness the by-products of a NPP, it is necessary to understand what implications the construction of a NPP would bring, both on the human and the industrial development of the surrounding areas. For example, it is important to know the minimum distance at which people can be settled, the allowed density, and whether there are legislations that *a priori* exclude the development of certain applications, as the related facilities interacting with the NPP are considered too hazardous. The analysis starts from the risk zoning around a NPP and the site selection criteria for its construction. Risk zoning is the identification of diverse EPZs (Emergency Planning Zones) around the NPP, where particular safety measures have to be taken: this chapter gives the official definition of EPZs and describes the risk zones around a NPP, according standards set by the IAEA (International Atomic Energy Agency) and the NUREG (NUclear REGulation of the USA), in order to supply a regulatory frame. Then, it describes the current status of EPZs in many countries. Thus, it notice that, despite suggestions provided by international legislations, each country can decide its own EPZ features, and the risk zoning methods vary significantly from country to country: international regulations are rather guidelines that can be taken as reference. It goes trough the site selection criteria in order to identify constraints and rules for the siting of a NPP and for population and industrial development in the adjacent areas. The main idea that emerged is that the presence of a particular facility cannot be excluded *a priori*: international standards state that although a facility may be regarded as potentially hazardous, its feasibility has to be investigated and justified through economic, environmental, safety and technical factors. It is also necessary to analyze its interaction with the NPP in order to definitively prove its safety.

2. SECTION ONE: THE EPZ

2.1. IAEA's Definition of EPZ

The IAEA (IAEA, 2003) defines three threat categories of nuclear reactors reported in Table 1.

Table 1. Emergency Planning Categories (IAEA, 2003)

Threat Category	Description
I	Facilities, such as NPPs, for which on-site events (including very low probability events) are postulated that could give rise to severe deterministic health effects off the site, or for which such events have occurred in similar facilities.
II	Facilities, such as some types of research reactors, for which on-site events are postulated that could give rise to radiation doses to people off the site that warrant urgent protective actions in accordance with international standards, or for which such events have occurred in similar facilities.
III	Facilities, such as industrial irradiation facilities, for which on-site events are postulated that could give rise to radiation doses that warrant or contamination that warrants urgent protective actions on the site, or for which such events have occurred in similar facilities.

Table 2. Threat categories for nuclear power reactors (IAEA, 2003)

Reactor power	Threat summary	Typical threat catH.
> 100 MWth	Off site: Emergencies involving severe core damage have the potential for causing severe deterministic health effects, including deaths. Radiation doses in excess of the urgent GILs (Generic Intervention Levels) are possible more than 5 km from the facility. Deposition resulting in radiation doses in excess of the relocation GILs and ingestion GALs (Generic Action Levels) is possible at great distances from the facility. An emergency not involving core damage has only a small potential for exceeding urgent GILs. On site: For core damage emergencies, doses sufficient to result in severe deterministic health effects, including deaths, are possible.	I or II
2 – 100 MWth	Off site: Radiation doses due to inhalation of short lived iodine in excess of urgent GILs are possible if cooling of the core is lost (core melt). On site: Potential for radiation doses in excess of urgent GILs if fuel cooling is lost. If shielding is lost, direct shine dose could exceed urgent GILs or result in severe deterministic health effects.	II or III
< 2 MWth	Off site: No potential for radiation doses in excess of urgent GILs. On site: Potential for radiation doses in excess of urgent GILs from inhalation (depending on design) if fuel cooling is lost. If shielding is lost, direct shine dose could exceed urgent GILs or result in severe deterministic health effects.	III

For most accident types, emergency response takes place over two distinct areas:

1. On-site area: it is the area surrounding the facility and within the security perimeter, fence or other designated property marker. This area is under the immediate control of the facility or operator;

2. Off-site area: it is the area beyond the on-site area. For facilities with the potential for emergencies resulting in major off-site releases or exposures (threat categories I and II), the level of planning will vary depending on the distance from the facility, as explained later.

The threat category of nuclear reactors depends on their power, as shown in Table 2.

Reactors with a power greater than 100 MWt (i.e. 33 MWe) will account for almost the totality of the market. Such designs include large reactors (LR) such as EPR Areva or AP 1000 by Westinghouse (about 4.500 MWth), or even small-medium reactors (SMR) like IRIS (335MWe - 1.000 MWth LWR). Moreover, all the reactors currently used in electro-nuclear power plants belong to category I. Thus, the chapter will only take into account EPZs for category I facilities. Emergency planning (EP) for category I plants is the most demanding. According to the IAEA, planning and implementing the capabilities to handle emergencies in category I facilities will ensure that the capability exists to handle events belonging to the other categories. However, for on-site and local organizations, planning and implementation should be based on local practices and activities (IAEA, 1997). As concerns off-site facilities, emergency planning can be discussed for two EPZs, as illustrated in Figure 3. and described as follows (IAEA, 2003).

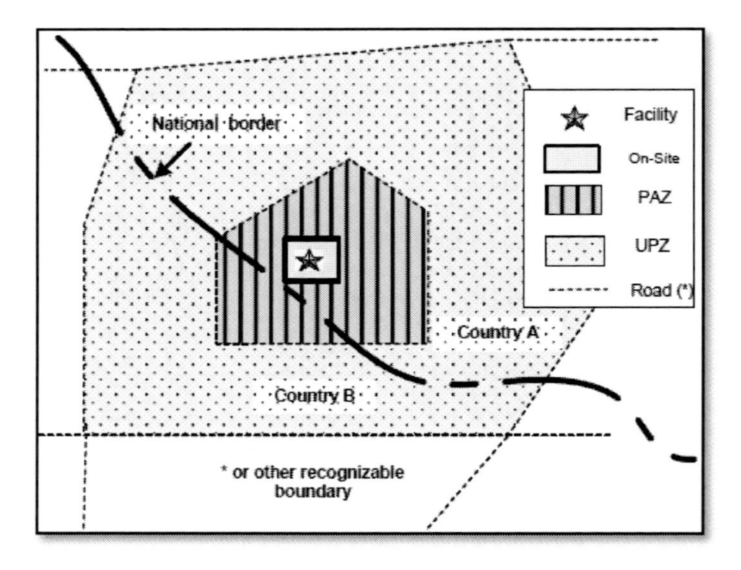

Figure 3. Concept of Emergency Zone (IAEA, 2003).

2.1.1. Precautionary Action Zone (PAZ)

This is a pre-designated area around a facility in threat category I, where urgent protective actions has been pre-planned and will be implemented immediately upon declaration of a general emergency. The goal is to substantially reduce the risk of severe deterministic health effects by taking protective action within this zone before or shortly after a release.

2.1.2. Urgent Protective Action Planning Zone (UPZ)

This is a pre-designated area around a facility in threat category I or II. Inside the UPZ, preparations are made to promptly implement urgent protective actions based on environmental monitoring data and assessment of facility conditions, the goal being to avert radiation doses specified in international standards.

2.1.3. Long Term Protective Action Planning Zone (LPZ)

It is the furthest pre-designated area around a facility and includes the UPZ. It is the area where preparations for the effective implementation of protective actions to reduce the long-term radiation dose from deposition and ingestion should be developed in advance (IAEA, 1997). When the IAEA-TECDOC 953 was updated in 2003, LPZ was replaced with "*Food Restriction Zone*", as shown by comparing Figures 4. and 5. As pointed out later, only the PAZ and the UPZ as belonging to the EPZ, excluding the LPZ (or FRZ), because it does not impose evacuation planning.

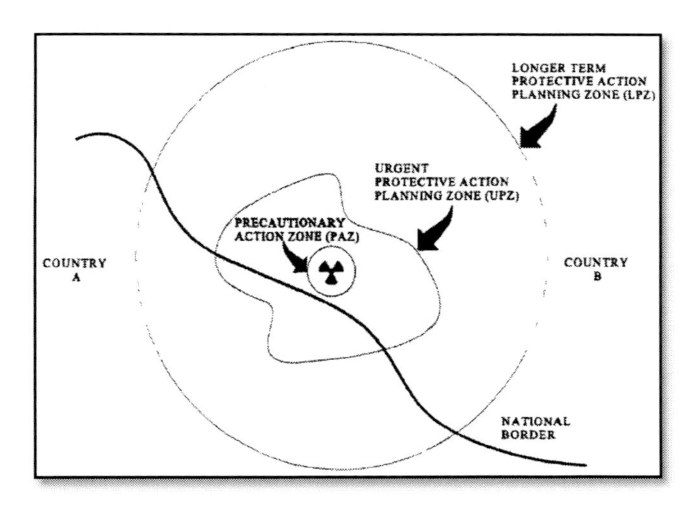

Figure 4. Emergency planning zones and radii (IAEA, 1997).

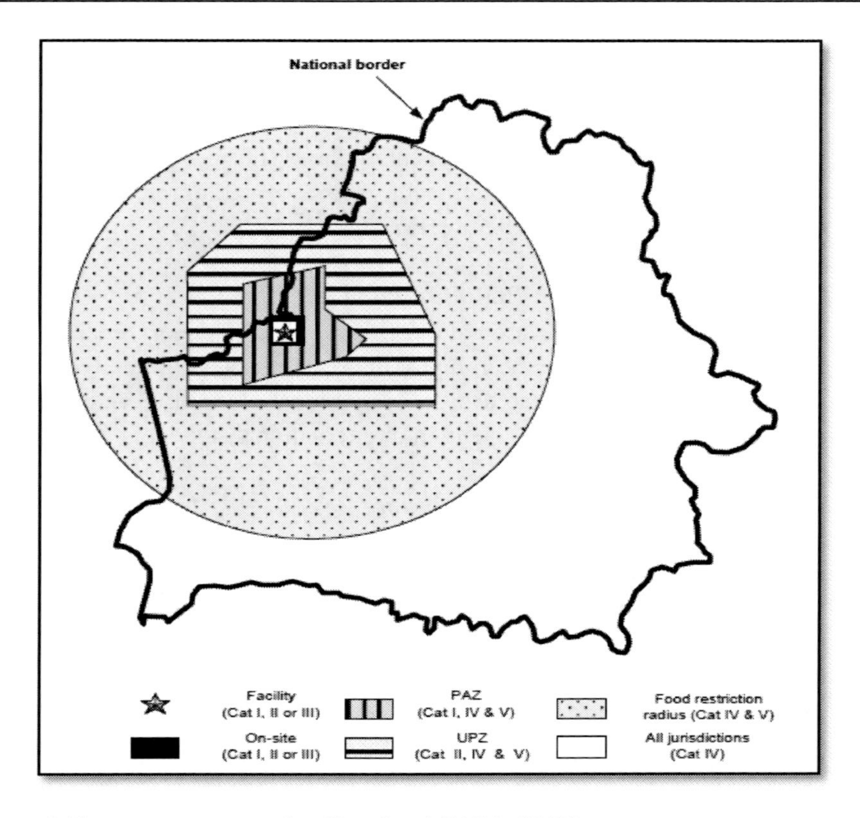

Figure 5. Emergency zones and radii updated (IAEA, 2003).

EPZs should be roughly circular areas around the facility, their boundaries defined by local landmarks (e.g. roads or rivers) to allow easy identification during a response. It is important to note that the zones do not stop at national borders. The size of the zones has to be determined by an analysis of the potential consequences following an accident. However, previous studies ((NRC, 1990) and (NRC, 1988)) of a full range of radiological and nuclear accidents provide a basis for generic zone sizes, as summarized in Table 3. It must be noticed that these suggestions are provided with recognition of the great uncertainties involved and variation by a factor of two or more during application is reasonable. The choice of the suggested radii represents a judgment of the distance to which it is reasonable to make advanced arrangements in order to ensure effective response. In a particular emergency zone, protective actions may be warranted only in a small part of the zones. For the worst possible emergencies, protective actions might need to be taken beyond the suggested radii (IAEA, 2003).

Table 3. Suggested emergency zones and radii for threat category I and II (adapted from (IAEA, 2003) and (IAEA, 1997))

Facility category	Reactors power	PAZ Radius	UPZ Radius	LPZ Radius (*)	FRZ Radius
I	> 1000 MWth	3-5 km	25 km	50-100 km	300 km
I	100-1000 MWth	0.5-3 km	5-25 km	50-100 km	50-300 km

Notes: (*) LPZ has been substituted with FRZ (Food Restriction Zone).

The suggested sizes for the PAZ were based on expert judgment considering the following (IAEA, 2003):

- urgent protective actions taken before or shortly after a release within this radius will prevent radiation doses above the early death thresholds for the vast majority of severe emergencies postulated for these facilities;
- urgent protective actions taken before or shortly after a release within this radius will avert radiation doses;
- radiation dose rates that could have been fatal within a few hours were observed at these distances during the Chernobyl accident;
- the maximum reasonable radius for the PAZ is assumed to be 5 km because:
    - except for the most severe emergencies, it is the limit to which early deaths are postulated;
    - it provides about a factor of ten reduction in the radiation dose compared to the dose on the site;
    - it is very unlikely that urgent protective actions will be warranted at a significant distance beyond this radial distance;
    - it is considered the practical limit of the distance to which substantial sheltering or evacuation can be promptly implemented before or shortly after a release;
    - implementing precautionary urgent protective actions to a larger radius may reduce the effectiveness of the action for people near the site, who are at the greatest risk.

The suggested sizes for the UPZ are based on expert judgment considering the following (IAEA, 2003):

- these are the radial distances at which monitoring to locate and evacuate hot spots (deposition) within hours/days may be warranted in order to significantly reduce the risk of early deaths for the worst emergencies postulated for power reactors;
- at these radial distances there is a reduction in concentration (and thus risk) by a factor of 10 of a radiation release, compared to the concentration at the PAZ boundary;
- this distance provides a substantial base for expansion of response efforts;
- 25 km is assumed to be the practical limit for the radial distance within which to conduct monitoring and implement appropriate urgent protective actions within a few hours or days. Attempting to conduct initial monitoring to a larger radius may reduce the effectiveness of the protective actions for the people near the site, who are at the greatest risk;
- under average meteorological (dilution) conditions, for most postulated severe emergencies, the total effective radiation dose for an individual beyond this radius would not exceed the urgent protective actions for evacuation.

This chapter reports only the information about the suggested sizes of PAZ and UPZ distances, and not those about LPZ and FRZ, because only the PAZ and the UPZ impose safety measures such as evacuation, that limit the presence of people thus posing impediments on the development of applications that require a high density of personnel/users at a short distance from the plant. Even though this chapter considers the PAZ and the UPZ, the radius of the zone that imposes evacuation can reach 25 km. Anyway, these distances are just a suggestions, and they are probably the results of very conservative criteria. It will be proved in section 1.3.1. that such a long distance for evacuation is taken into consideration by very few countries.

3. Definition of EPZ by NRC

NRC (Nuclear Regulatory Commission, USA) developed the most important reference documents for risk zoning around a NPP and emergency planning zones (NRC, 1998; NRC, 2003; NRC, 1998). According to these documents, the zones around a NPP are the following (Figure 6).

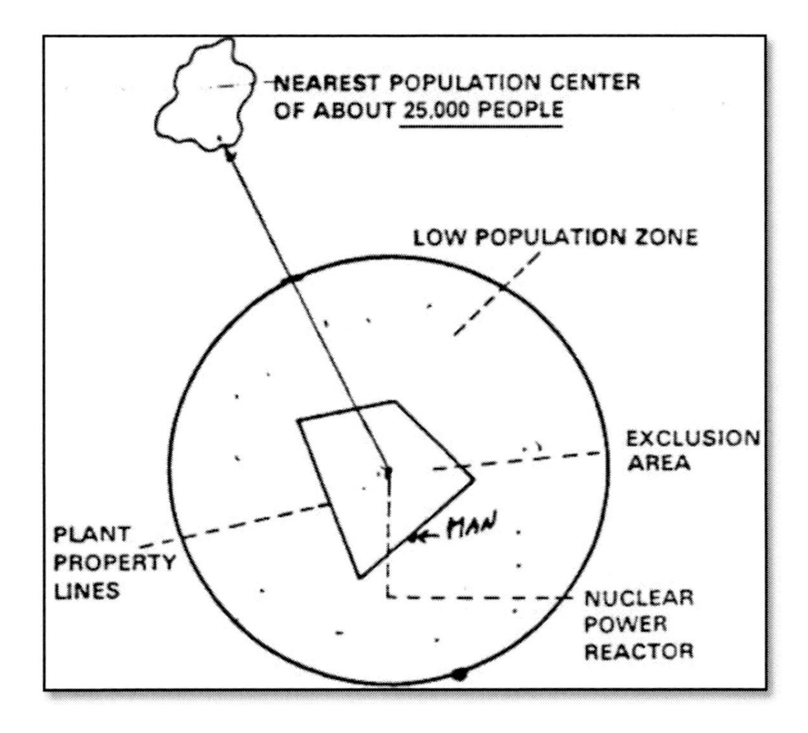

Figure 6. 10CFR100 requirements (NRC, 1962).

3.1. Exclusion Area (EA or EAB)

It is the area surrounding the reactor, where the reactor licensee retains the authority to determine all activities, including exclusion or removal of personnel and property from the area. This area could be traversed by a highway, railroad, or waterway, if they do not interfere with normal operations of the facility and it is possible to control traffic on the highway, railroad, or waterway, in case of emergency, to protect the public health and safety. Residence within the exclusion area shall normally be prohibited. In any event, residents shall be subject to ready removal in case of necessity. Activities unrelated to operation of the reactor may be permitted in an exclusion area under appropriate limitations, provided that no significant hazards to the public health and safety will result (NRC, 2003).

The EA size is not fixed: it must be of such size that an individual located at any point on its boundary for two hours immediately following onset of the postulated fission product release would not receive a total radiation dose to the whole body in excess of 25 rem or a total radiation dose in excess of 300 rem to the thyroid from iodine exposure. Thus, the required EA size involves consideration of the atmospheric characteristics of the site as well as plant design (NRC, 2003).

The concept of Exclusion Area originated in the USA in the early 1950s, when an acute awareness existed about the potential effects of nuclear accidents on the nearby population. This idea was mooted primarily to insulate the public from the harmful effects of low-probability, high-consequence accidents. The earliest attempt to size the EA (the so called "rule of thumb") was made by NRC (NRC, 1950): the exclusion distance was numerically specified as a circle of radius $R = 0.01 \times P^{0.5}$ [miles], where P is the reactor thermal power (kW). This formula would not yield practical sizes for medium-sized or large power reactors: for a typical 3000 MWth reactor, this formulation gives an exclusion radius of 17.3 miles (27.9 km). Thus, the US siting practice as embodied in 10 CFR Part 100 for the determination of the exclusion boundary and the low population zone around a reactor, updated in 2003 (NRC, 2003), lately defined these radial distances in a more correct way, based on the radiation dose after an accident. The methodology for implementing this in the US Context is coded in the NRC document TID-14844 (NRC, 1962).

When implemented, the exclusion distances for most US reactors fall in the range of 0.5–1.6 km.

The factors determining the exclusion boundary are: reactor type and power, engineered safety features, containment design and characteristics of the site. The US code of practice assumes a severe beyond design basis accident and does not give credit to design features save the containment (BARC, 1975).

Some examples of Exclusion Area Boundaries (EAB) recently assessed for different reactors are the following:

1. ABWR (Lungmen nuclear project, Taiwan, expected to be commissioned in July 2010): in the Preliminary Safety Analysis Report it is stated "the distance from the centre of reactor building to the EAB is 300 m. There are no waterways, railroad, or public highways that traverse the boundary of the exclusion area"(AEC, 2005);
2. CANDU: "because of the lower design leak rate from containment, the EAB radius for the siting of CANDU 9 can be as small as 500 m, significantly reducing site area requirements for CANDU 9 plants. This is an important advantage in the context of meeting siting requirements and land availability" (Hedges, 2005);

3. EPR: "Site boundary considerations for new nuclear Darlington (Canada)" considers ACR-1000, EPR and AP-1000 reactors. EPR meet the dose acceptance criteria from RD-337 with an EAB of 500 m (OPG, 2009).

3.2. Low Population Zone (LPZ)

It is the area immediately surrounding the exclusion area which contains residents, the total number and density of which are such that there is a reasonable probability that appropriate protective measures could be taken in their behalf in the event of a serious accident. These guides do not specify a permissible population density or total population within this zone because the situation may vary from case to case. Whether a specific number of people can, for example, be evacuated from a specific area, or instructed to take shelter, on a timely basis will depend on many factors such as location, number and size of highways, scope and extent of advance planning, and actual distribution of residents within the area.

The LPZ size is not fixed. It must be of such size that:

1. an individual located at any point on its outer boundary who is exposed to the radioactive cloud resulting from the postulated fission product release (during the entire period of its passage) would not receive a total radiation dose to the whole body in excess of 25 rem or a total radiation dose in excess of 300 rem to the thyroid from iodine exposure;
2. the population centre distance (distance from the reactor to the nearest boundary of a densely populated centre containing more than about 25,000 residents) is at least one and one-third times the distance from the reactor to the outer boundary of the LPZ.

The boundary of the population centre should be determined considering population distribution, not political boundaries. Where very large cities are involved, a greater distance may be necessary because of total integrated population dose consideration. The size of the LPZ depends upon atmospheric dispersion characteristics and population characteristics of the site, as well as aspects of plant design. (NRC, 2003).

For plants licensed in USA in the 1960s and early 1970s a LPZ radius of about 5 km was found acceptable. (BARC, 1975)

The TID-14844 (NRC, 1962) provides the distances needed for the exclusion area, the LPZ and the population centre as a function of the thermal power of the LWR to be sited at a particular location. These distances are recapped in Figures 7. and 8.

3.3. Plume Exposure Pathway and Ingestion Exposure Pathway Zones

To facilitate a pre-planned strategy for protective actions during a radiological emergency, there are two emergency planning zones around each NPP (NRC, 1998). The exact size and shape of each zone is a result of detailed planning which includes consideration of the specific conditions at each site, unique geographical features of the area, and demographic information. This pre-planned strategy for an emergency planning zone provides a substantial basis to support activity beyond the planning zone in the extremely unlikely event it would be needed. The two zones are described as follows and in Table 4:

1. Plume Exposure Pathway zone (PEP): the PEP zone has a radius of about 16 km (10 miles) from the reactor site. Predetermined protective action plans are in place for this zone and are designed to avoid or reduce radiation dose from potential exposure of radioactive materials. These actions include sheltering, evacuation, and the use of potassium iodide where appropriate. The principal exposure sources from these pathways are:

   a) whole body external exposure to gamma radiation from the plume and from deposited materials;
   b) inhalation exposure from the passing radioactive plume.

   The duration of principal potential exposures could range in length from hours to days. Figure 9. depicts a typical 10-mile PEP zone map. The centre of the map is the location of the commercial NPP reactor building. Concentric circles of 2, 5, and 10 miles have been drawn and divided into triangular sectors identified by letters from A to R. Municipalities identified to be within the 10-mile PEP have been assigned numbers from 1 to 24. The triangular sectors provide a method of identifying which municipalities are affected by the radioactive plume as it travels;

2. Ingestion Exposure Pathway zone (IEP): the IEP has a radius of about 50 miles (80 km) from the reactor site. Predetermined protective action plans are in place for this zone and are designed to avoid or reduce radiation doses from potential ingestion of radioactive materials. These actions include a ban of contaminated food and water. The principal exposure from this pathway would be from ingestion of contaminated water or foods such as milk or fresh vegetables. The duration of principal exposures could range in length from hours to months.

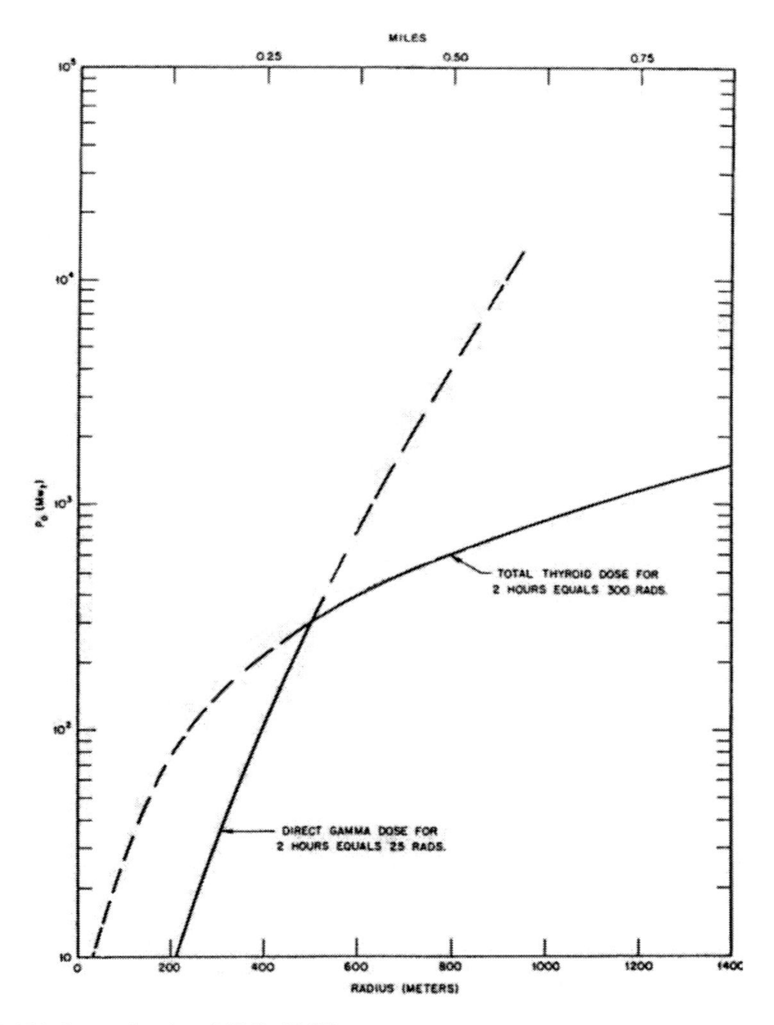

Figure 7. EA determination (NRC, 1962).

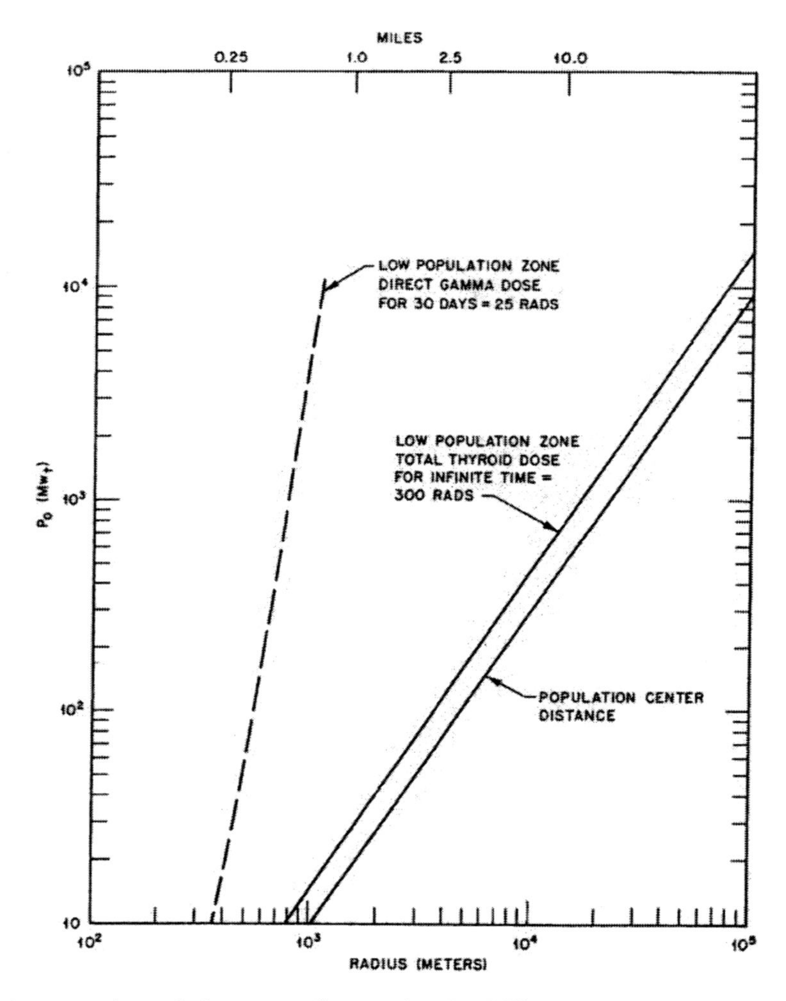

Figure 8. LPZ and population centre distance (NRC, 1962).

Table 4. Guidance on size of PEP and IEP zones (NRC, 1998)

Accident phase	Critical organ and exposure pathway	EPZ radius
PEP	Whole body (external radiation) Thyroid (inhalation) Other organs (inhalation)	About 16 km
IEP	Thyroid, whole body, bone marrow (ingestion)	About 80 km

Figure 9. Typical 10-mile PEP zone map (NRC, 1998).

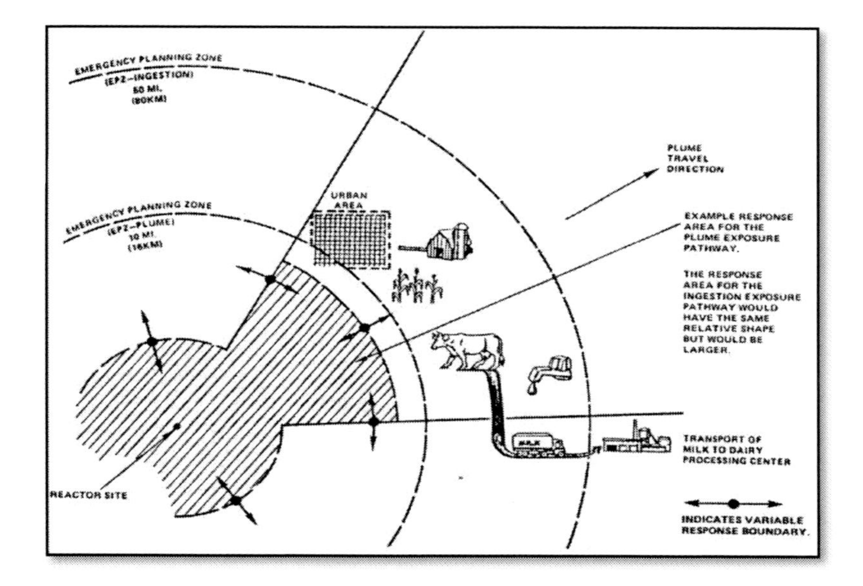

Figure 10. Concept of PEP and IEP zones (NRC, 1998).

4. EPZs WORLDWIDE

4.1. Differences in EPZ Regulations around the World

Current regulations vary among IAEA Member States. They either prescribe the EPZ size through a deterministic, or a risk-based approach, or appear as some combination thereof. The technical basis is not always clearly spelled out (IAEA, 2005).

Around nuclear installations, planning zones for the implementation of countermeasures are pre-established, but their sizes vary among different countries:

- the planning zone for evacuation is, in general, in the order of 10 km around the nuclear installation;
- the planning zones for sheltering and stable iodine are generally of the same size, and range from 10-20 km, larger than the evacuation zones. Choosing identical planning zones indicates that sheltering and stable iodine are often implemented together.

In all cases, zone sizes are based on detailed analyses of possible accidents, their severity and consequences (OECD/NEA, 2003).

In Table 5, an overview of the current practices for risk zoning around NPPs is presented.

Table 5. Overview of emergency planning practices in different countries (Kirchsteiger, 2006) and (OECD/NEA, 2003)

Country	Zones
Australia	Zone 1: 500 m pre-planned evacuation zone Zone 2: 2.2 km (dependant upon conditions) ANSTO exclusion zone – 1.6 km
Belgium	Evacuation: 10 km
Canada	Evacuation zone: 7 km Sheltering zone: 10 km Iodine zone: 10 km
Czech Republic	NPP Dukovany: 10 km evacuation zone, 20 km sheltering and stable iodine zone NPP Temelin: 5 km evacuation zone, 13 km sheltering and stable iodine zone
Finland	Protective zone: 5 km distance from the facility EPZ: extending to about 20 km from the facility

Country	Zones		
France	Evacuation: 5 km EPZ (sheltering and iodine): 10 km		
Germany	Central Zone: Surrounds the nuclear facility in a 2 km radius. Intermediate Zone: A circle with a radius of up to about 10 km around the NPP Outer Zone: A circle with a radius of up to about 25 km around the NPP		
Hungary	Internal zone: 3 km Sheltering zone, where evacuation can be considered: 31 km Zone where sheltering can be considered: 71 km		
Japan	Sheltering zone, including evacuation zone (for NPPs): 8 to 10 km		
Luxembourg	Iodine: up to 25 km Evacuation and sheltering: case by case decision		
Netherlands	Radius Implementation zone around the NPP:		
	<100 MWe: 5 km	100-500 MWe: 10 km	>500 MWe: 15 km
	Radius Countermeasure zones for the respective MWe, distance from the NPP:		
	Evacuation		
	0	5	5
	Iodine prophylaxis		
	4	10	15
	Sheltering		
	7	20	30
	In a segment depending on the wind direction. For evacuation > 100 MWe always also in a circle with 2 km radius		
Norway	For two research reactors, zones are being established according to the draft IAEA Safety Series on emergency planning and response		
Country	Zones		
Slovakia	Internal zone: 3 km for Bohunice Inner emergency zone: up to 12-15 km in radius around the NPP Indication zone: up to approximately 50 km in radius around the NPP EPZ: 30 km Bohunice, 20 km Mochovche (divided into zones of 5 and 10 km)		
South Africa	Internal zone: 5 km UPZ: 5 to 16 km LPZ: 80 km		
Sweden	Inner emergency zone: up to 12-15 km in radius around the NPP Indication zone: up to approximately 50 km in radius around the NPP		
Switzerland	Internal zone: 3 to 5 km Zone 1: Approximately 4 km in radius around the NPP (= sheltering zone) Zone 2: Approximately 20 km in radius (= sheltering zone)		
United Kingdom	1 to 3 km		
USA	PEP Zone: 16 km IEP Zone: 80 km		

Notes: (*) Internal zone is generally defined as the zone in which no further development is allowed.

In a recent paper the current status of the emergency and risk zones around a NPP has been analyzed for several countries. It pointed out that (Kirchsteiger, 2006):

- many countries use the relevant IAEA documents (e.g. the 2003 updated version of IAEA TECDOC 953);
- there are significant differences in the EPZ radii in different countries, ranging from a few up to 80 km, as shown in Table 5;
- there is a striking contrast in the extent of using probabilistic information to define EPZs between the nuclear and other high risk industry sectors, such as the chemical process industry, and the reasons for these differences are not entirely clear, since the risk in the chemical industry is similar to that of the nuclear sector;
- the approach to emergency planning is, in general, strongly deterministic. The usual approach is that a reference accident is defined and used as a basis for drawing up the emergency plans;
- the difference seems to be more related to risk perception than to actual risk potential;
- there is a strong need to communicate risk information to the public both before and following an accident, and to educate the public so they can understand risk information in a comparative sense; the issue needs to be addressed on whether there are any advantages or disadvantages in imposing larger EZs.

4.2. EPZ Reduction and the Role of Small Reactors

The EPZ itself does not pose particular issues to the co-location of NPPs and other facilities, it imposes some limitations on the presence of people; reducing the EPZ would reduce this problem. According to INPRO (*International project on innovative nuclear reactors and fuel*) and GIF (*Generation IV International Forum*), innovative, small reactors could allow the reduction or even elimination of the EPZ. This section deals with the main issues linked to the presence of the EPZ, the motivations and goals for its reduction and the main advantages this would bring.

Emergency planning requirements may represent a significant burden for the plant owner (utility), both in the construction and in the operating phases. During the construction, it may be necessary to build infrastructures (highways) to comply with the requirements. During operating phases, it is necessary to maintain an evacuation capability in a relatively wide area. Moreover, one of the consequences of emergency planning requirements is the "freezing" of any human development in a large area around the plant. Finally, the fact that the off-site zone around a NPP is subject to particular constraints may spread distrust towards nuclear power safety (Augutis, 2005).

4.2.1. Attempts to Reduce the EPZ

Even though the concept of EPZ has been joined with nuclear power since the very beginning, many attempts to reduce it have been experimented (IAEA, 2006):

1. in 1985, the licensee of the plant of Calvert Cliffs (Maryland) requested an EPZ reduction from ten to two miles, and in 1986 the plant of Seabrook (Texas) requested its reduction to one mile. Both petitions were rejected by the NRC: the former because severe accident issues were still under study by the NRC and the latter because the supporting documentation did not contain sufficient justification. After these two early failures, there were no more licensee petitions, but rather studies and investigations continued, performed by various organizations, fuelled by the excellent safety record of operating plants and the enhanced safety characteristics of advanced reactors;
2. in 1993, the NRC staff raised the following issue: "should advanced reactors with passive advanced design safety features be able to reduce EPZ and requirements?". No changes were actually proposed, but it indicated that a revision of the EPZ was not impossible;
3. in 1997, an evaluation of emergency planning for advanced reactors was conducted by the NRC in SECY-97-020, reaching the conclusion that the existing NUREG-0396 approach was also appropriate for the new plants, that were on the drawing boards. At the same time, however, it was recognized that "changes to emergency planning requirements might be warranted to account for the lower probability of severe accidents and the longer time period between accident initiation and release of radioactive material for most severe accidents associated with evolutionary and passive advanced LWRs". In order to justify these types of changes, three main issues had to be addressed:

    1) Probability level below which accidents will not be considered for emergency planning (the so-called "cut-off probability");

2) Use of increased safety in one level of defence in depth to justify reducing requirements in another level;
3) Acceptance by federal, state and local authorities.

4) the task of Group 1 within the CRP i25001 (Coordinated Research Project on small reactors without on-site refuelling) is to develop a methodology and to identify regulatory approaches to revise (reduce or eliminate) off-site emergency measures such as evacuation and relocation for NPPs with innovative reactors. The general objective of Group 1 activities assumes there may be several equivalent, similar, or related practical implementations, such as to:

- eliminate the need for off-site response;
- revise the need for off-site relocation and evacuation measures;
- reduce the size of the EPZ;
- reduce the EPZ to fit within site limits (eliminating the off-site response).

It is recognized that while the complete elimination of off-site EPZ may be difficult, eliminating or even reducing most costly measures may provide similar economic effects/benefits (IAEA, 2005).

4.2.2. EPZ Reduction/Elimination Goals

The trend is to improve the level of safety for future NPPs (Generation IV reactors and small/medium reactors - SMR). This would significantly reduce the probability of severe accidents and releases of radioactive material from the plant. In principle, this could be considered to reduce, or perhaps eliminate, the need for emergency planning. Further considerations need to be given about how EP and EPZs may be defined for future NPPs where the risk in term of large off-site releases of radioactivity was much lower than in current plants. Consideration needs to be given on whether the moral obligation to provide an EP would outweigh the technical conclusion that EP would not be required any more. (Kirchsteiger, 2006) The idea of EPZ reduction for advanced nuclear reactors is based on two factors (Maioli, 2006):

1. the safety level of new reactors: for example, in the case of IRIS, a LERF (Large Early Release Frequency) of 10^{-9} has been estimated;
2. the fact that EP is based on risk perception rather than on a risk assessment.

Elimination of EPZ is one of the goals of INPRO (International Project on Innovative Nuclear Reactors and Fuel Cycles) and GIF (Generation IV International Forum):

- IAEA TECDOC 1434 states that "innovative nuclear energy systems (INS) shall not need relocation or evacuation measures outside the plant site, apart from those generic emergency measures developed for any industrial facility", which means that INS could be sited in very similar locations to those of other energy producing systems. The corresponding criterion is specified as "probability of large release of radioactive materials to the environment", and the acceptance limit considered is $<10^{-6}$ per plant-year, or excluded by design,
- it also suggests that the end point should be to make the risk of INS comparable to that of industrial facilities used for similar purposes, so that for INS there will be no need for relocation or evacuation measures outside the plant site;
- one of the goals of the GIF is to reach a condition with "no need for offsite response". A reasonable measure of this goal could be expressed as "no credible accident scenarios that could result in offsite release of radiation exceeding US protection action guidelines. These guidelines may change as improved radiation dose-response models are developed" (IAEA, 2005).

Achieving licensing without EPZs would offer significant societal and economic benefits to member countries, general public and plant owners/operators, including ((Augutis, 2005) and (IAEA, 2006)):

- no a priori impediment to further development and settlements in areas around the plant;
- increased public acceptance of nuclear power, since NPPs would be treated as any other industrial facility;
- reduced need for infrastructure to facilitate rapid evacuation, thus reducing connected costs;
- reduced operational costs, since there would be no need for special training of personnel and for periodic evacuation drills;
- enabling of co-generative applications, including district heating, desalination, industrial process heat supply, where the plant cannot be

located remotely from the intended user (cost of extended transmission lines avoided);

- enabling the choice of sites that would reduce transmission costs;
- enabling a wider choice of sites in countries with relatively high population density.

4.2.3. Correlation between Size and EPZ Dimension

There is a correlation between the reactor size and the EPZ size. First of all, it must be underlined that the reactor size varies from country to country: some countries rely predominantly on large reactors, other countries on small ones, and finally some countries have a balanced mix. The situation for the countries analyzed is summarized in the following table.

It is possible to divide EPZ sizes in three categories, as shown in table 8. The connection between the reactor size and the EPZ size for various countries is provided in table 8 and figure 12. The tendency is to establish large EPZs if LRs are employed and small EPZs for small reactors. There are no countries using small reactors that have large EPZ, but there are some (France and Germany) that have small EPZ even though all reactors are large.

Table 7. Countries and reactor size adopted

Reactor size	Countries
Predominantly large (more than 700 MW)	Belgium, France, Germany, South Africa, USA
Predominantly small (less than 700 MW)	Hungary, Netherlands, Slovakia, UK
Mixed (small and large in the same proportion)	Czech Republic, Finland, Japan, Switzerland, Canada

Table 8. EPZ size and relative countries

EPZ size	Radius range	Countries
Small	Less than 5 km	Slovakia, Hungary, UK, Switzerland, France, Germany
Large	Between 5 and 10 km	Netherlands, Finland, Canada, Czech Republic, South Africa
Very Large	More than 10 km	USA, Japan, Belgium

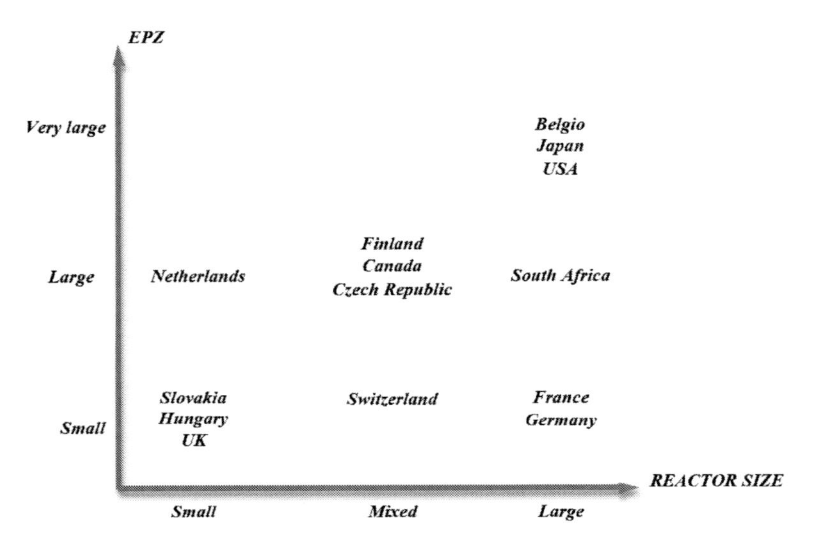

Figure 12. Correlation between reactor size and EPZ size.

CONCLUSION

The collected data demonstrate that:

- EPZs size around a NPP are neither set by an international regulation, nor imposed by the reactor vendor or by other authorities;
- the exact size and configuration of the EPZs should be determined with respect to local emergency response needs and capabilities, as they are affected by conditions as: atmospheric characteristics of the site, plant design, demography, topography, land characteristics, access routes, jurisdictional boundaries;
- agencies like the IAEA and the NUREG tried to suggest typical sizes for EPZs, but all the suggestions are provided with recognition of the great uncertainties involved (a variation by a factor of two or more during application is reasonable) and in any case exact sizes must be confirmed by case-specific studies;
- current EPZs are extremely different from country to country;
- the determination the radius of an EPZ is related to the size (in terms of power) and to the level of safety of the reactor, but it has to be the result of a precise and complete case-by-case risk assessment analysis;

- the guides do not specify a precise permissible population density or total population within the closest zone to the NPP because the situation may vary from case to case: whether a specific number of people can be evacuated from a specific area on a timely basis will depend on many factors such as location, number and size of highways, as well as actual distribution of residents within the area;
- the importance of EPZ reduction (in terms of off-site emergency planning elimination) for innovative small reactors with enhanced safety has been recognized, and it is based on the fact that:

    - it would lower the transmission cost for co-generative applications (district heating);
    - it would enable a wider choice of sites to locate NPP;
    - it would eliminate a priori impediments for the economic and human development in the area surrounding the plant;

- EPZ is based on risk perception rather than on risk assessment.

5. Section Two: Potential Applications in the EPZ

5.1. Introduction

The EPZ represents an inhibition to the urban and economic growth of the area within its borders; however, despite hindering the development of the territory as a whole, can also represents an attractive opportunity to single appliances: all the by-products of a NPP are available inside the EPZ perimeter, and the EPZ itself can be regarded as a valuable by-product, because its wide, uninhabited, low cost areas are the ideal location for many industrial/energetic facilities.

This fact raises two major questions:

1. What synergies do the nuclear by-products offer inside an EPZ?
2. What industrial/energy applications could be implemented to exploit these synergies?

This section tries to answer these two questions, identifying attractive synergies between nuclear power and different kinds of applications. In general co-generation is the simultaneous generation of heat and electricity; when a heat source is used to produce only electricity (e.g., through a steam turbine), about one third of the heat is converted, while the remaining two thirds are wasted. Part of this heat can be recovered extracting a certain amount of steam from the turbine. When the heat source is a nuclear reactor the issue is called nuclear cogeneration. Depending on the temperature reached in the reactor, nuclear co-generation applications can be at low temperature or at high temperature. It is wise to divide the applications in two main group: low temperature and high temperature.

5.2. Low Temperature Applications

Low temperature applications presented in the following paragraphs are actual or viable in the short-term (5 years or less), thanks to the maturity of the employed technologies and to the commercial availability of nuclear reactors capable of providing heat at the needed temperature. Considerable experience has been accumulated worldwide both for nuclear-powered district heating and for industrial uses of nuclear heat that will be now briefly described.

5.2.1. Nuclear District Heating

In district heating the steam extracted from high and/or low-pressure turbines is fed to heat exchangers to produce hot water/steam, which is delivered to the consumers. Depending on the transportation distance and the number of end users, a certain number of pumping stations are located between the heating source and end users. Heat transportation pipelines are installed either above or under-ground. They are well insulated, in order to minimize heat losses. Steam from low-pressure turbines is usually used for the base heat load, while steam from high-pressure turbines is used, when needed, to meet the peak heat demand. The portion of steam retrieved for heat production represents a part of the total steam produced by the reactor, the remaining portion of the steam being used to produce electricity (IAEA, 2002). In principle, any portion of the heat can be extracted from co-generation reactors as district heat, subject to design limitations. Co-generation plants, when forming part of large industrial complexes, can be readily integrated into an electrical grid system to supply any surplus generated. In turn, they would serve as a back-up for the assurance of the energy supply. This guarantees a high degree of flexibility (IAEA, 2007).

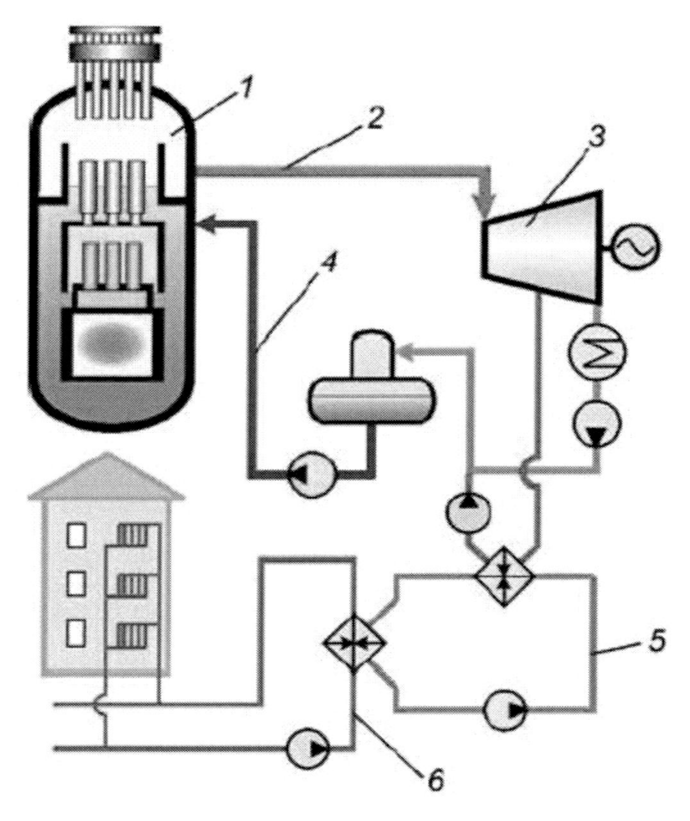

Figure 14. Nuclear district heating concept. Elaboration from (Kutznetsov, 2008).

Correct function of interface equipment is an important basis for good operating performance. Operating experiences of interface equipment for nuclear district heating are not different from those in commercial thermal plants (except for the radioactivity monitoring devices) (IAEA, 1998).

Figure 14. depicts a simplified scheme of nuclear district heating. Its principal components are the nuclear reactor (1), the supply of steam to the turbine (2), the turbine unit (3), the supply of feed-water to the reactor (4), the exchange line (5) and the heat consumer (6).

5.2.2. Technical Requirements for Nuclear District Heating

Required Temperatures

District heating systems are supplied with steam or hot water in a typical temperature range of 80-150 °C (IAEA, 2007).

Suitable Reactors

From the technical point of view, nuclear reactors are basically heat-generating devices. There is plenty of experience of using nuclear heat in district heating, so the technical aspects can be considered well proven. There are no technical impediments to the application of nuclear reactors as the heat source for district heating.

In principle, any type and size of nuclear reactor can be used for these purposes. Thus, all existing reactor types (light water, heavy water, fast breeder, gas cooled and high temperature) are potentially applicable to cogeneration for district heating (IAEA, 2007).

Distance from the User

Due to high losses over longer transmission distances, the heat source must be relatively close to the customer, typically within 10-15 km (IAEA, 2007).

In commercial scale heating networks, the transportation distances are usually less than 10 km, in most cases between 3 and 6 km (IAEA, 1998). Anyway, in some cases the heat source can be located further from the customer (up to 100 km) depending on the economics based on the size of the plant and the level of insulation technology (OECD/NEA, 2004).

Heat losses along the network can be extremely reduced if pre-insulated pipes are utilized: a typical result is a 3% loss on the transported power (0.1°C/km, if the temperature difference between feed and return is 15°C along a 5 km network). The maximum loss is about 1°C/km. (RENAEL, 2004).

The impact of the distance on heat transportation cost is given in table 10, where the cost of a 5-km-transfer is taken as a base. The distance between the nuclear power plant and the user is not a problem in terms of heat losses, but the cost of heat transportation grows linearly with the distance: thus, it should be minimized in order to reduce transmission costs.

Table 10. Impact of distance on the heat transportation costs (IAEA, 2002)

Distance [km]	Cost of heat transportation
5	1
10	2.5-3.5
15	4.5-5.5
20	6.5-8.0

Capacity

The district heat generation capacities are determined by the collective demands of the customers. In large cities an installed capacity of 600-1200 MWth is normal, while the demand is much lower in towns and small communities (10 to 50 MWth). Large capacities of 3000-4000 MWth are exceptional (IAEA, 2007).

Load Factor

The annual load factor is normally not higher than 50%, since heat is supplied only in the colder part of the year. This is still way below what is needed for base load operation of plants. (IAEA, 2007). The annual load factor can increase if the distribution of sanitary hot water is provided.

Expected Availability of a Heat Distribution Network

District heat involves the supply of space heating and hot water through a district heating system, which consists of heat plants (producing electricity simultaneously) and a network of distribution and return pipes. Thus, the availability of a heat distribution network plays an important role in the prospect of nuclear district heating development. (IAEA, 2007)

Availability Factor

The experience shows that availability factors of 70%, 80% or even 90% can be achieved (similar to the availabilities achieved by fossil fuelled power plants). The frequency and duration of unplanned outages can be kept very low with good preventive and predictive maintenance, but not eliminated: consequently, redundancy is needed. Multiple-unit co-generation power plants, modular design, or backup heat sources are necessary to achieve the required availabilities. (IAEA, 2007)

Backup Capacity

To ensure a reliable supply of heat to the residences served by the district heat network, adequate backup heat generating capacity is required. This implies the need for redundancy and generating unit sizes: at least two nuclear power units, or a combination of nuclear and fossil fired units, corresponding to only a fraction of the overall peak load (Csik, 1997).

Heat Storage

Heat storage allows a matching of the heat supply to the heat demand. Today there are many examples of short-term storage, for instance, on the daily scale that relies on hot water accumulator tanks. In the future, more innovative concepts for long-term storage facilities may be realized, such as storage in underground water layers (IAEA, 2007).

Safety

Potential radioactive contamination of the district heating networks is avoided by appropriate measures. No incident involving radioactive contamination has ever been reported for any of the reactors used for these purposes (Csik, 1997). Because of the need to site the source close to the customer, nuclear safety is very important. It is not only required that the level of safety is technically sufficient, it is also necessary that the adequacy of safety be sufficiently proved to the public and confirmed by the licensing process. (IAEA, 2002).

Conclusion

- All existing reactor types are potentially applicable to cogeneration with district heating purposes, and several European countries already have experience in nuclear district heating for residential, agricultural and commercial sector: thus, nuclear district heating is technically feasible;
- Nuclear district heating can compete economically in densely populated areas with individual heating arrangements. Economic studies generally indicate that district heating costs from nuclear power are in the same range as costs associated with fossil-fuelled plants, but a site-specific comparison of the cost of nuclear heat production with those of competing technologies is necessary;
- Nuclear district heating offers the possibility of strongly reducing air pollution in urban areas: the full integration of external costs in the nuclear case would render nuclear district heating the most attractive option in economic terms, even compared with renewable;
- There is a major trade-off in siting reactors intended for district heating: the site must satisfy both the requirements of the nuclear plant (the EPZs require the location of district heating users far from the reactor) and of the heat application (low transmission costs are achieved if users are located near the reactor);
- The heat output of a large reactor is far larger than the demands likely for district heating;

- The development of nuclear district heating will be favoured by the diffusion of small, modular reactors: low cost, better match of the heat demand, enhanced safety, potential to reduce EPZ and increase social acceptance.

5.3. Nuclear Process Heat

Process heat implies the supply of heat required for industrial processes from several centralized heat generation sites through a steam transportation network. Wasted heat from the nuclear reactors can be used for this purpose: from a technical point of view, the functioning is similar to that of district heating. Thus, most considerations done for district heating are valid here. Differences come from the required temperatures and the annual load factor, which are both higher.

5.3.1. Technical Requirements for Nuclear Process Heat

Required Temperatures

Within the industrial sector, process heat is used for a very large variety of applications with different heat requirements and with temperature ranges covering a wide spectrum. The application of nuclear industrial process heat is tightly connected to the temperature (Csik, 1997):

- The lower range, up to about 200 to 300 °C includes industries such as seawater desalination, pulp and paper, or textiles;
- Chemical industries, oil refining, oil shale and sand reprocessing, and coal gasification are examples of industries with temperature requirements of up to the 500 to 600 °C level;
- Refinement of coal and lignite, and hydrogen production by water splitting are among applications that are renewing the interest and they require temperatures between 600 and 1000 °C;
- The upper range above 1000 °C is dominated by the iron/steel industry.

This section considers only the applications that are feasible with commercially developed reactors: that means up to 600 °C (low and medium temperature).

High temperature applications are discussed in section 0. A series of industrial processes at low and medium temperature and their temperature ranges are represented in figure 16.

Suitable Reactors

The required heat parameters determine the applicability of different reactor types. There are no technical impediments to the application of nuclear reactors as heat sources for process heating, thus, all existing reactor types and sizes are potentially applicable to producing process heat depending on the required temperature of the processes (IAEA, 2002).

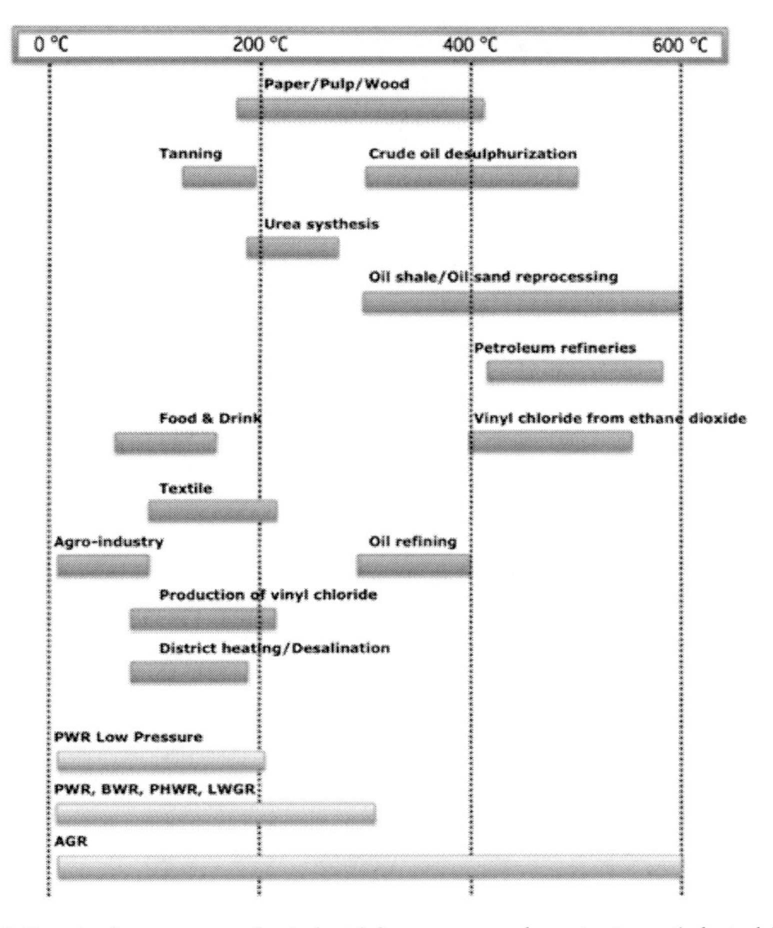

Figure 16. Required temperature for industrial processes and reactor types (adapted from (IAEA, 2007), (IAEA, 1998) and (IAEA, 2002).

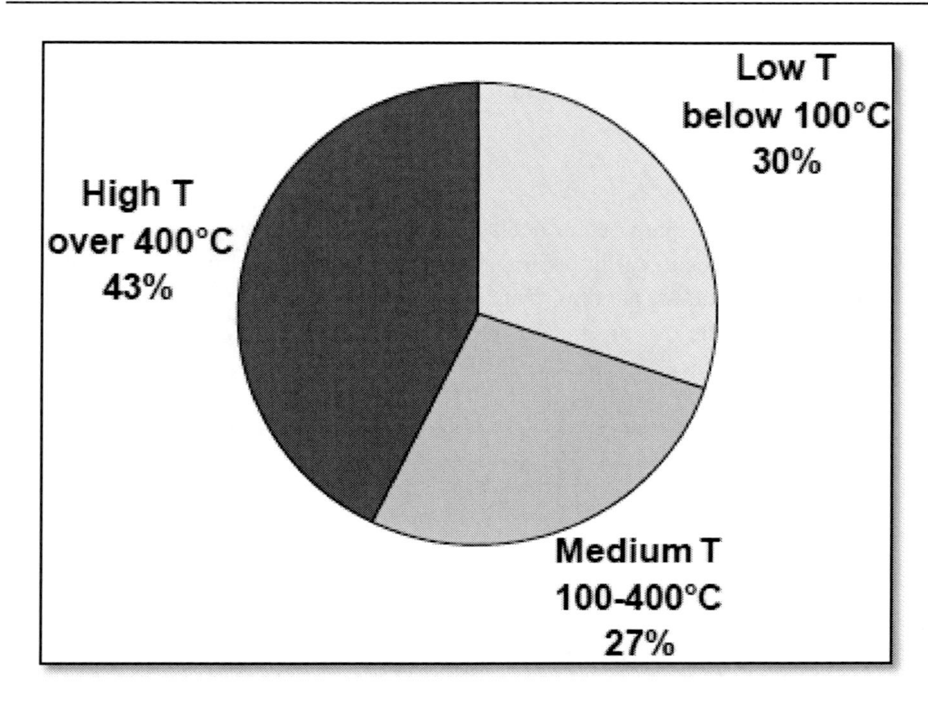

Figure 17. Share of industrial heat demand by temperature in EU (EUROHEATCOOL, 2006).

The applications of temperature ranges between 20° C and 600 °C and the reactors meeting these requirements are represented in Figure 16. However, an important market for nuclear process heat at low temperature exists. As illustrated in figure 17, about 30% of the total industrial heat demand is required at temperatures below 100°C and 57% at temperatures below 400°C.

Moreover, in several industrial sectors, such as food, wine and beverage, transport equipment, machinery, textile, pulp and paper, the share of heat demand at low and medium temperature (below 250°C) is about, or even above, 60% of the total figure (ECOHEATCOOL, 2006).

Distance from the Customer

Due to thermal losses over transportation, the heat source has to be relatively close to the customer (IAEA, 2007). Although, this is not as critical as it is in district heating systems: since industrial complex are not densely populated (on the contrary of residential areas), they can be sited more easily near the NPP in order to optimize the trade off safety/transmission costs. This would also lead to a better exploitation of the EPZ.

Annual Load Factor

Since process heat demand does not depend on climatic conditions, the supply of industrial heat is more uniform throughout the year than that of district heat.

The demands of large industrial users usually have base load characteristics, with annual load factor of 70-90%. Nuclear reactors, which are typically run in base load operation, will be quite useful in this context (IAEA, 2007).

Needed Availability Factor

Almost all industrial users need the assurance of energy supply with a very high degree of reliability and availability, approaching 100% in particular for large industrial installations and energy intensive processes. The average adequate steam supply availabilities for chemical processing and oil refineries are respectively 98% and 92% (IAEA, 2007).

Backup Capacity

Most industrial processes require highly reliable heat supply, even though some processes (e.g. drying) can also work with interruptible heat supply. Industrial heat consumers can be supplied with steam from a multi-unit or from a single unit nuclear station.

In both cases, one or several devices for the backup capacity are required. The frequency and duration of unplanned outages can be kept very low with good preventive and predictive maintenance.

Availability and reliability of a reactor, however, can never reach the nearly 100% levels required by most large heat users: multiple unit co-generation power plants, modular designs, or backup heat sources are suitable solutions for redundancy (IAEA, 2007).

Safety

Potential radioactive contamination of the networks is avoided by appropriate precautions, such as intermediate heat transport circuits with pressure gradients, which act as effective barriers.

No incident involving radioactive contamination has ever been reported for any of the reactors used for these purposes (Csik, 1997). The siting of an industrial heat user close to the NPP will require specific safety features appropriate to the location and the application (IAEA, 2000).

5.3.2. Market for Nuclear Process Heat

Market Fragmentation

The industrial heat market is highly fragmented, and it is characterized by a steady decrease in the number of users as the power requirements become higher (IAEA, 2000):

- about half of the users require less than 10 MWth;
- another 40% of the users require between 10 and 50 MWth;
- about 99% of the users are included in the range of less than 300 MWth, which account for about 80% of the total energy consumed;
- individual large users with energy intensive industrial processes cover the remaining portion of the industrial heat market with requirements up to 1000 MWth, and exceptionally even more.

Thus, the large-scale introduction of heat distribution system supplied from a centralized nuclear heat source need the presence or the development a sort of industrial park, where several users are concentrated.

Process Heat Users: Main Industries

Generally, the industries that are main consumers of heat are:

- Petroleum and coal processing;
- Chemical and fertilizers;
- Primary metal;
- Paper and products;
- Food and products.

The apportionment varies from country to country, but the chemical and petroleum industries are the largest consumers worldwide. These would be key target clients for possible applications of nuclear energy (IAEA, 2002).

Market size does not matter for nuclear penetration. The main question is whether nuclear technologies can prove to be competitive. The market for industrial heat is highly competitive. Heat is produced predominantly from fossil fuels, with which nuclear energy will have to compete (IAEA, 2002).

Worldwide Experiences in Nuclear Process Heat

There is experience in providing process heat for industrial purposes with nuclear energy in Canada, Germany, Norway, Switzerland and India. New plants are being designed in Russia, the Republic of Korea and Canada (IAEA, 2007). The most significant examples of nuclear process heat are listed in Table 13. Both for the number of different users served and for the huge quantity of thermal power supplied, the most synergic plant is the Bruce Energy Centre in Canada, where steam is used for heavy water production plants and for an adjacent industrial park. It is the world's largest nuclear steam/electricity generating complex. It includes eight CANDU nuclear reactors with a total output of over 7.200 MWe, the world's largest heavy water plant. The initial development focused primarily on agriculture-based industry. Then, a sustainable development model was presented, with the aim of demonstrate commercial application of "closed loop" and integrated systems, the introduction of nuclear hydrogen and absorption of CO_2. The sustainable development model is based on the following points (IAEA, 2000):

- Cogeneration of electricity and process steam using a nuclear reactor;
- A menu of feedstocks ranging from farm produced carbo-hydrates and solid wastes to low grade carbon sources and carbon dioxide;
- A series of state of the art processing, synthesizing and refining processes;
- End products that have markets and in their own right have environmental value-added.

Table 13. Experiences in industrial process heat applications (adapted from (IAEA, 2007))

Country (Location)	Reactor type	Start of reactors operation	Power [MWe]	Heat delivery [MWht]	T at interface [°C]	Remarks
Canada (Bruce)	CANDU	1981	848 (8)	5.350		D_2O production and six industrial heat customers
Germany (Stade)	PWR	1983	640 (1)	30	190/100	Salt refinery
Switzerland (Goesgen)	PWR	1979	970 (1)	45	220/100	Cardboard factory
India (Kota)	CANDU	1980	160 (1)	85	250	D_2O

The six private industries currently established in the park are (IAEA, 2007): a plastic film manufacturer, a 30.000 mq greenhouse, a 12 million liter/year ethanol plant, a 200.000 ton/year alfalfa dehydration, cubing and pelletising plant, an apple juice concentration plant and an agricultural research facility.

Siting and Construction

Similar to nuclear district heating, the close siting of a nuclear plant to the customer is preferable, as the heat transportation costs grow significantly with distance. Respect to residential complexes, industrial process heat users do not have to be located within highly populated areas. Many of the process heat users, in particular the large ones, can be, and usually are, located outside urban areas, often at considerable distances. This makes joint siting of nuclear reactors and industrial users of process heat not only viable, but also desirable in order to drastically reduce the heat transport costs, provided that the co-siting does not adversely affect the safety case for the nuclear installation (IAEA, 2000). In Germany and Switzerland there have been experiences with nuclear process heat and the distances from the industries were respectively 1.5 km (the PWR of Stade for a salt refinery) and 2 km (the Goesgen PWR for a cardboard factory) (OECD/NEA, 2004). Installing a new nuclear co-generation plant close to existing and interested industrial users has better prospects. Even better would be a joint project whereby both the nuclear co-generation plant and the industrial installation requiring process heat are planned, designed, built and operated together as an integrated complex (IAEA, 2007).

The Role of SMRs in Nuclear Process Heat

Coupling a large reactor with a small industrial facility does not allow a significant exploitation of heat from the reactor. The only chance to use a relevant fraction of the available heat from a large reactor is a large industrial complex requiring a high quantity of steam for different businesses (e.g. Bruce Eco Industrial Park in Canada, see section *Worldwide experiences in nuclear process heat*). Moreover, the EPZ around a NPP could be so large that the location of a lot of industries is not only viable, but also preferable in order to exploit this unused area. Such a kind of multi-business industrial park is quite difficult to implement, as it requires an extremely accurate choice of businesses and the presence of interested investors.

The reasoning could be inverted as well: if a high demand of heat is difficult to find, it is possible to reduce the offer. In this sense, the diffusion of small, innovative reactors with lower power and less EPZ requirements, could increase the attractiveness of coupling the nuclear power plant with a small industrial user.

According to this, the development of nuclear process heat applications could depend on the development of SMRs. For large size reactors used in co-generation mode, electricity will be the main product. Such plants, therefore, have to be integrated into the electrical grid system and optimized for electricity production. For reactors in the SMR size range, and in particular for small and very small reactors, the share of process heat generation would be larger, and heat could even be the predominant product. This would affect the plant optimization criteria, and could present much more attractive conditions to the potential process heat user. Consequently, the prospects of SMRs as co-generation plants supplying electricity and process heat are considerably better than those of large reactors (IAEA, 2007).

Conclusion

- All existing reactor types and sizes are potentially applicable to producing process heat, depending on the required temperature of the processes;
- Process heat has base load characteristics, as well as nuclear reactors: the matching between demand and supply is better than in the district heating case;
- The siting issue is not as critical as it is for district heating, because industrial complexes do not require high population density and they can be located near the NPP (i.e. inside the EPZ). This would lead to a better exploitation of the EPZ;
- The industrial process heat market is highly fragmented (few large users, lot of small users) and it is difficult to find such demanding users that can harness a significant amount of the heat supplied by a large reactor. Thus, there are two options to favour the utilization of nuclear heat for industrial processes:

    1. The concentration of small industrial users in so-called industrial parks to match the demand and the supply: if the interaction between NPP and other plants is proven to be safe, they can be located inside the EPZ;
    2. The large-scale commercialization of small reactors;

In the first case, a joint project, whereby both the nuclear co-generation plant and the industrial installations requiring process heat are planned, designed, built and operated together is preferable.

5.4. High Temperature Applications

The feasibility of high-temperature applications exploiting nuclear heat is dependent upon the commercialization of nuclear reactors operating at adequate temperature, which is envisaged in about 10-30 years, depending on the technology (WNA, 2009).

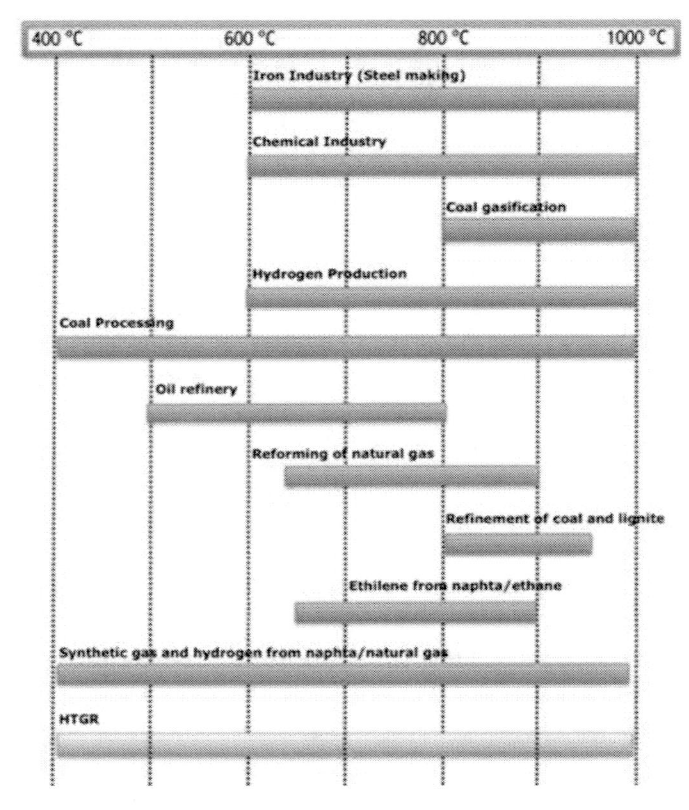

Figure 19.Temperature required by some industrial processes at high temperature (adapted from (IAEA, 2007), (IAEA, 1998) and (IAEA, 2002).

Table 14. Gasifier operating temperatures (Minchener, 2005)

Gasifier configuration	Operating temperature [°C]
Moving bed gasifier	1000
Fluidized bed gasifier	900
Entrained flow gasifier	1200-1600

The technology of some high-temperature processes discussed in this section is not yet mature as well (e.g. thermo-chemical H_2 production), despite their advanced stage of development and the confidence of international literature. For these reasons high-temperature applications are characterized by a higher level of uncertainty than low-temperature ones. High temperature applications are divided into traditional ones (process heat at high temperature) and innovative ones. Traditional applications will not be discussed in depth, as information given in the previous section about low temperature process heat apply to them as well. The only difference lies in the temperature, therefore in the reactors and, as a consequence, in the technology availability, which is supposed to be due in about 2030. Our choice is to give relevance to innovative applications. All high temperature reactors are small, innovative reactors of Generation IV, thus it makes sense to hypothesize a small EPZ for the applications.

Considerations referred to low temperature process heat are similar to high temperature traditional applications. Here, the development of the applications is bound to the development of HTGR (High Temperature Gas Reactors). In order to show the potential use of high temperature reactors for industrial steam supply, the following figure presents the different temperatures required by some typical industrial processes.

Table 15. Potential biomass gasifier feedstock and heating value (Ciferno and Marano, 2002)

Biomass	Heating value [MJ/Kg]
Agricultural residues	
Sawdust	19,3
Bagasse	17
Corn cob	17
Short rotation woody crops	
Beech wood	18,4
Herbaceous energy crops	
Switchgrass	15,4
Straw	17,0
Miscanthus	12,0
Municipal solid waste	
Dry sewage	8,0
Coals	
Subbituminous	24,6
Bituminous	27,0

5.4.1. Gasification via Nuclear Heat

Gasification is a means to convert fossil fuels, biomass and wastes into either a combustible gas or a synthesis gas for subsequent utilization (Minchener, 2005), consisting primarily of hydrogen (H_2) and carbon monoxide (CO).

The oxidant used can be air, pure oxygen, steam or a mixture of these gases (Ciferno and Marano, 2002). Under an economic point of view, the gasification process converts solid or liquid feedstock of lesser market value than premium gas or liquid fuels, into a synthesis gas that is suitable for use in electricity production or for the manufacture of chemicals, hydrogen, or transportation fuels (Stiegel and Maxwell, 2001).

Technological Options

There are three gasifier configurations, described below: they differ in flow geometry and in process parameters such as temperature and pressure.

1. Moving bed gasifiers (also called fixed bed g.) have been the traditional choice for gasification. Gases flow relatively slowly upward through the bed of feedstock material (Minchener, 2005); depending on the direction of the airflow, moving bed gasifiers are further classified as updraft, downdraft or cross-flow, and each class is characterized by different operating temperatures (McKendry, 2002) (see table 14);
2. Fluidized bed gasifiers, in which feedstock particles are suspended in the gas flow, and the material entering the gasifier is mixed with that already undergoing gasification. Two main kinds of fluidized bed gasifiers are in use: circulating fluidized bed g. and bubbling bed g. (McKendry, 2002) (see table 14).
3. Entrained flow gasifiers, in which pulverized coal particles and gases flow concurrently at high speed. They are the most commonly used gasifiers for coal gasification (Minchener, 2005), but the need for a finely divided feed material (<0.1–0.4 mm) creates problems for fibrous materials such as wood, thus making the process unsuitable for most biomass materials (McKendry, 2002).

All gasifier configurations require air, oxygen or steam at high temperatures (McKendry, 2002) (see Table 14): this prevents their combining with state-of-the-art/LWR NPPs, that cannot reach, the needed temperatures.

Feedstock

Gasification can be operated either with biomass or coal as a feedstock, each having different characteristics, availability and costs:

1. Gasification of biomass: biomass is the organic material from recently living things, including plant matter from trees, grasses, and agricultural crops (Ciferno and Marano, 2002). The chemical composition of biomass varies among species, but basically consists of high moisture content, a fibrous structure consisting of lignin, carbohydrates or sugars, and ash-Biomass possesses a heating value lower than that of coal (see Table 15), and it is very non-homogeneous in its natural state: this non-homogeneous character poses difficulties in maintaining constant feed rates to gasification units, often resulting in a low heating value for the product syngas, typically <2.5 MJ/m3 (Ciferno and Marano, 2002); to be considered interchangeable with conventional fossil fuels and to ensure maximum flexibility for industrial or utility applications, the syngas heating value needs to be above 11 MJ/m3 (the heating value for natural gas being approximately 37 MJ/m3) (Turn, 1999).
2. Gasification of coal: coal gasification involves converting solid coal into a gaseous fuel that can be used similarly to natural gas; the objective of the conversion is to mitigate some of the drawbacks associated with the combustion of solid coal (WNA, 2010). In particular, gasification allows a significant reduction of air emissions from the direct combustion of coal (e.g. particulates, sulphur oxides and heavy metals). An important advantage of coal gasification is that high reserve availability all over the world. In general, the use of gasified coal has the same advantages as the use of natural gas, but the world current reserves of coal are much larger than those of natural gas;
3. Co-gasification of coal and biomass: biomass, whether as a dedicated crop or a waste-derived material, is renewable. However, the availability of a continuous biomass supply can be problematic (for example, crop supply may be decreased by poor weather or by alternative uses, and the availability of a waste material can fluctuate depending on variations in people's behavior) (Komabe, Hanaoka, and Fujimoto, 2007). The principle of co-gasification is to adjust the amount of coal fed to the gasifier so as to alleviate biomass feedstock fluctuations. Co-gasification is a new area of study, and only pilot studies are being carried on.

Products and Applications

Different outputs of the gasification process are listed and described below:

1. *Gasification can create Substitute Natural Gas (SNG)* from coal or other feedstock: using a “methanation" reaction, the SNG - chiefly carbon monoxide (CO) and hydrogen (H_2) - can be then profitably converted to methane (CH_4) (Mozaffarian, Zwart, Boerrigter, and Deurwaarder, 2004);
2. *Gasification can generate power directly:* gasification can produce electric power via a direct combustion boiler/steam turbine: this system has a low efficiency (between 20 and 25%) (Ciferno and Marano, 2002). Power generation can also be accomplished via gasification of biomass, followed by a combustion engine, combustion turbine, steam turbine or fuel cell. These systems can produce both heat and power and can achieve greater system efficiencies, in the range of 30 to 40%. If the feedstock is coal, the *Integrated Gasification Combined Cycle* (IGCC) is the baseline choice: this particular coal-to-power technology allows the continued use of coal without the high level of air emissions associated with conventional coal-burning technologies. In contrast, conventional coal combustion technologies capture the pollutants after combustion, which requires cleaning a much larger volume of the exhaust gas, leading to increased costs, reduced reliability, and generating large volumes of sulfur-laden wastes that have to be disposed (Minchener, 2005);
3. *Gasification can synthesize chemicals and fertilizers:* it produces valuable byproducts such as ammonia and phosphates, that have potential on the fertilizer market (Ro, Cantrell, Elliott, and Hunt, 2007);
4. *Gasification can produce H_2* for the hydrogen economy: production of H_2 from renewable sources derived from agricultural or other waste streams offers the possibility to lower greenhouse gas emissions (without carbon sequestration technologies) (Levin and Chahine, 2009). The key problem with gasification is how to separate and purify the H_2 from other gases in the syngas; the technology is not yet mature for a satisfying implementation.

5.4.2. Hydrogen Production via Nuclear Heat

As an alternative path to the current fossil fuel economy, a hydrogen economy is envisaged in which hydrogen would play a major role in energy systems and serve all sectors of the economy, substituting for fossil fuels (IAEA, 2007).

Hydrogen possesses a number of attractive features that could allow it to become a key secondary energy carrier in the future:

- Hydrogen combustion (either hot or cold) is generally clean, since it does not produce the characteristic emissions of fossil fuel combustion. The problem of NO_x production from high temperature combustion is practically eliminated in modern engine designs (Conte, Iacobazzi, Ronchetti, and Vellone, 2001).
- Technologies similar to those used for the combustion of fossil fuels can be used for hydrogen combustion to generate heat, electricity and propulsion energy; for example, hydrogen can be used as fuel in catalytic combustions (in diffusion burners, fuel cells), in internal combustion engines and in gas turbines (WNA, 2010).
- Hydrogen is storable, which is convenient for an energy carrier and gives the possibility of making the energy system much more flexible than at present, in particular by using the conversion of electricity to hydrogen (through water electrolysis) and vice versa (through fuel cells), as necessary (WNA, 2010).
- Hydrogen could be a third product from power plants, in addition to electricity and heat (Forsberg, 2003).

Making the fullest possible use of the above advantages, hydrogen can be considered a key element of an environmentally benign and sustainable energy system, including transportation.

Market Perspectives

The annual world consumption of H_2 is about 50 million tons, which is used primarily for ammonia production and conversion of heavier crude oils to clean liquid fuels (Forsberg, 2003).

The hydrogen market has been growing steadily in the last decade, and this growth is expected to continue with a 10% yearly rate (Blanchette, 2007), doubling the demand by 2020.

Moreover, in the long term, if the hydrogen economy occurs, the use of hydrogen for all our transportation needs would require 18 times more hydrogen than the ammount currently used. The usage of hydrogen for all our non-electric energy needs would imply 40 time more the “normal ammount” (Schultz, Brown, Besenbruch, and Hamilton, 2003).

Hydrogen Production Methods

Nuclear energy provides a source of heat to produce H_2. Multiple processes are being investigated to produce H_2 from water and heat. If nuclear energy is to be used for H_2 production, the nuclear reactor must deliver heat at conditions that match the requirements imposed by the H_2 production process. The viability of H_2 production from nuclear power ultimately depends upon the economics, which, in turn, depend upon both the proposed methods of H_2 production and the available reactors. Four methods have been proposed to produce H_2 from nuclear power:

- Electrolysis: electrolysis of water to produce H_2 is an old technology that is used today to produce ultrapure H_2 and to produce H_2 in small quantities at dispersed sites. Electrolysis is not currently competitive for the large-scale production of H_2 (Forsberg, 2003).
- Steam reforming: today, H_2 is produced primarily from the steam reforming of natural gas. Steam reforming is an energy-intensive endothermic low-pressure process requiring high-temperature heat as an input. Natural gas is used as the reduced chemical source of H_2 and burned to produce heat to drive the process at temperatures of up to 900°C. The amount of natural gas required for steam reforming can be significantly reduced when heat is provided by a nuclear reactor. The Japan Atomic Energy Research Institute is currently preparing to demonstrate the production of H_2 by steam reforming of natural gas with the heat input provided by its High-Temperature Engineering Test Reactor (HTTR). The nuclear power plant provides heat that replaces that from a gas flame. Because this system uses standard H_2 production technology, it represents the near-term nuclear H_2 technology, once HTTRs are commercially viable (Forsberg, 2003).
- Hot electrolysis: electrolysis can be operated at high temperatures (700–900°C) and low pressures to replace some of the electrical inputs with thermal energy. Because heat is cheaper than electricity, the H_2 costs via this production method could ultimately be lower than those for traditional electrolysis. Equally important, the high temperature results in better chemical kinetics within the electrolyser that reduces equipment size and inefficiencies. However, the technology is at an early stage of development although it derives much of its technology from solid-oxide fuel cells. Hot electrolysis requires collocation of H_2 production close to the nuclear reactor to provide the heat (Forsberg, 2003).
- Thermo-chemical hydrogen production: hydrogen can be produced by direct thermo-chemical processes, in which the net reaction is: heat plus

water yields H_2 and oxygen. These are the leading long-term options for production of H_2 using nuclear energy. For low production costs, however, high temperatures (more than 750°C) are required to ensure rapid chemical kinetics (i.e., small plant size with low capital costs) and high conversion efficiencies. Of the advanced methods for hydrogen generation using nuclear power, thermo-chemical cycles have received the most attention because current estimates indicate that thermo-chemical H_2 production costs could be as low as 60% of those from room-temperature electrolysis (Forsberg, 2003).

- Biomass gasification: hydrogen can be produced with lower or no greenhouse emissions via the gasification of agricultural or other waste (Levin and Chahine, 2009).

Thermo-chemical processes are currently regarded as the most promising technology for massive production of hydrogen in the next decades (Forsberg, 2003).

Process Requirements

Process requirements for H_2 production via nuclear steam reforming of methane, hot electrolysis, and thermo-chemical cycles are similar. All three technologies impose similar requirements on the nuclear reactor (Forsberg, 2003):

- *Reactor power:* H_2 production facilities match better with reactor powers below 1000 MWe (Forsberg, 2003), but larger reactor scales (such as the 1650 MWe AREVA reactors that are planned to be constructed in Italy (WNA, 2009)) do not prevent H_2 production applications.
- *Peak temperature:* all the methods previously described (see Hydrogen production methods) but electrolysis requires high temperature heat (750–900°C).
- *Temperature range of delivered heat:* all of the endothermic high-temperature chemical reactions operate at a nearly constant temperature. Heat should therefore be delivered over a small temperature range.
- *Pressure:* the chemical reactions go to completion at low pressures. High pressures reverse the desired chemical reactions. The H_2-nuclear interface should be at low pressure to minimize the risk of pressurization of the chemical plant and minimize high-temperature materials strength requirements.

- *Isolation:* the nuclear and chemical facilities should be isolated from each other so that upsets in one facility do not impact the other. The system must also minimize tritium (radioactive hydrogen) production and transport from the reactor to the H_2 production facility.

Nuclear Reactor Selection

The high peak temperatures reached by all the processes (750-900°C, see *Process requirements*) except standard electrolysis are not endurable by currently commercialized reactors (WNA, 2009); therefore, although several methods to produce H_2 using high temperature heat are available, significant development work is required before any of these processes can be actually put in practice.

Sandia National Laboratories evaluated various nuclear reactors for their ability to provide the high temperature heat needed, and to be interfaced safely and economically to the hydrogen production process (Schultz, Brown, Besenbruch, and Hamilton, 2003). The recommended reactor technologies were supposed to require minimal development to meet the high temperature requirement and also to be free from any significant design, safety, operational or economic issues.

The following conclusions were drawn:

- PWR, BWR and organic-cooled reactors: not recommended, because they cannot achieve sufficiently high temperatures.
- Liquid-core and alkali metal-cooled reactors: they imply serious development risk, due to material concerns at the needed temperatures.
- Heavy metal and molten salt-cooled reactors: promising, but they require a significant development effort.
- Gas-core reactors: not recommended, too speculative at present.
- High-temperature gas-cooled reactors: baseline choice. In particular, only modest development is needed for helium gas-cooled reactor, which has historically been considered the one reactor that would be used for the purpose. Alternatively, a reactor can be designed specifically for H_2 production: the Advanced High-Temperature Reactor (AHTR) has been proposed; this concept is similar under many features (core design, fuel cycle) to the General Atomics modular helium reactor (Forsberg, 2003).

Economics of H_2 Production

Nuclear power plants are characterized by high capital costs and low operating costs; therefore, the economics are strongly dependent upon maintaining base-load operations with continuous output. Two characteristics of hydrogen help doing so:

- Constant base-load demand for H_2 favors technologies with low fuel costs, such as nuclear energy (Forsberg and Peddicord, 2001).
- Hydrogen packing (increasing the pressure) creates significant storage capacity, which can mitigate potential variations in demand; using the techniques developed by the natural gas industry, H_2 storage in large volumes is expected to be relatively low cost (Forsberg, 2003).

In addition, the need for security, the difficulty in finding social acceptance for nuclear plants and the economic advantages of using common facilities encourage siting multiple reactors at each site.

Economics of H_2 Distribution

Hydrogen transport is the major concern for the accomplishment of the hydrogen economy: if the hydrogen economy occurs as is prefigured, the scale of H_2 production is expected to evolve from distributed to midsize and only eventually (after 2030) to centralized.

Central station plants are assumed to have a production capacity of 1.200.000 kilograms per day (kg/d) and to operate with a 90 percent or higher capacity factor, therefore producing on average 1.080.000 kg/d H_2 and supporting nearly 2 million cars; midsize plants are assumed to have a production capacity of 24,000 kg/d (operating with a 90 percent capacity factor, they produce on average 21,600 kg/d H_2 which is enough to support about 40.000 cars); distributed plants have different production capacities corresponding to the differing capacity factors: those that operate with a 90 percent capacity factor are assumed to have a production capacity of 480 kg/d H_2, producing on average 432 kg/d (Committee on Alternatives and Strategies for Future Hydrogen Production and Use, 2004).

The nuclear energy source is compatible, for the number and size of plants, with centralized production (Committee on Alternatives and Strategies for Future Hydrogen Production and Use, 2004). If no breakthrough technologies are conceived, dedicated pipelines will be the most convenient solution for the transport of hydrogen from central station plants to users; line transmission of hydrogen, although, is expected to be highly capital-intensive, because costly steel and valve metal seal connections will be required in order to avoid long-term embrittlement and possibilities of leakage.

According to the analysis conducted by the Committee on Alternatives and Strategies for Future Hydrogen Production and Use, pipeline shipment and dispensing will cost \$0.96/kg H_2, which is essentially equal to the cost of H_2 production from natural gas, and higher than the cost of its production via thermal splitting with nuclear energy.

If and when extensive new hydrogen transmission pipelines are needed in the decades ahead, research in such areas as lower-cost pipeline materials, technology for dual-use of natural gas-and-hydrogen pipelines, layout optimization, and even pipeline emplacement technologies will be of critical importance.

5.4.3. Shale Oil Extraction

If carbon dioxide releases from liquid-fuels production are to be minimized, liquid fuels should be produced only from high-quality light crude oils; unfortunately, the resources of light crude oil are limited (Forsberg, 2009).

What is required is a technology to create large quantities of light crude oil without the release of large quantities of greenhouse gases: one option is a nuclear light-oil production system (Forsberg, 2008). This system may allow massive underground resources of fossil fuels, which are economically unrecoverable with existing technologies, to be converted into liquid fuels (Forsberg, 2009). Examples include the following:

- Old oil fields: over half the oil remains in a depleted oil field trapped by capillary forces between grains of sand or within cracks in the rock (Forsberg, 2009).
- Tar sands: tar sands are a mixture of sand, clay, water, and bitumen (viscous heavy oil). Unlike conventional oil, bitumen is too viscous to be pumped to the surface. The feasibility of oil recovery from tar sands is limited to surface deposits and underground deposits where steam heating can reduce the viscosity of the oil until it flows (Finan, Miu, and Kadak, 2005).
- Oil shale: oil shales are fine-grained sedimentary rocks containing relatively large amounts of organic matter (known as 'kerogen') from which significant amounts of shale oil and combustible gas can be extracted (World Energy Council, 2007); shale oil, when adequately processed, can be utilized as a crude oil substitute in most applications (Ots, 2007).
- Soft coal: soft coal, if heated, is converted to chat and a liquid fuel.

The extraction of oil from these categories of fossil deposits poses a challenge to the oil industry (Forsberg, 2009).

Shale Oil Reserves in the World

World shale oil resources[1] derivable from oil shale beds are estimated by the *European Academies Science Advisory Council* (EASAC) at approximately 3.2 trillion US barrels (EASAC, 2007). Two-thirds of the listed deposits are located in North America, while Europe accounts for approximately 12%.

The Russian territory holds more than 60% of European oil shale, and the Italian peninsula contains most of the remaining quantity (about 20% of the total, see Figure 20) (EASAC, 2007).

5.4.3.1. Shale Oil Extraction Technologies

The two classes of shale oil recovery are surface mining, which is the traditional means to extract shale oil from oil shale, and in-situ refining.

Surface Mining

Surface mining is operated through an open-pit recovery of oil shale with heavy-ton trucks and electric or hydraulic shovels; the ore is then sent to an extraction plant where the rock is separated from the kerogen, which undergoes a refining process (see Figure 21, left side: traditional refining) (Finan, Miu, and Kadak, 2005).

In-Situ Refining

Starting in the 1970s, researchers began to examine methods for underground oil recovery from the previously described fossil deposits; because of technological developments, concerns about greenhouse gas and CO_2 emissions, and higher oil prices, these technologies have now progressed to field testing, with initial leasing of properties for commercial production in pioneer countries with large oil shale reserves (Forsberg, 2008). The technology is conceptually simple (see Figure 21, right side: In-situ refining): a fossil deposit is heated to temperatures around 370°C through the injection of high-temperature heat at 700°C from the heater well; as the temperature increases, any volatile hydrocarbons will vaporize (be distilled), move as gases toward a recovery well, condenses in the surrounding cooler zones, and be pumped out of the ground as a liquid or vapor (Forsberg, 2009).

---

[1]Resources also comprehend those quantities of a commodity that are estimated to be potentially recoverable but which are not currently considered commercially recoverable (EASAC, 2007).

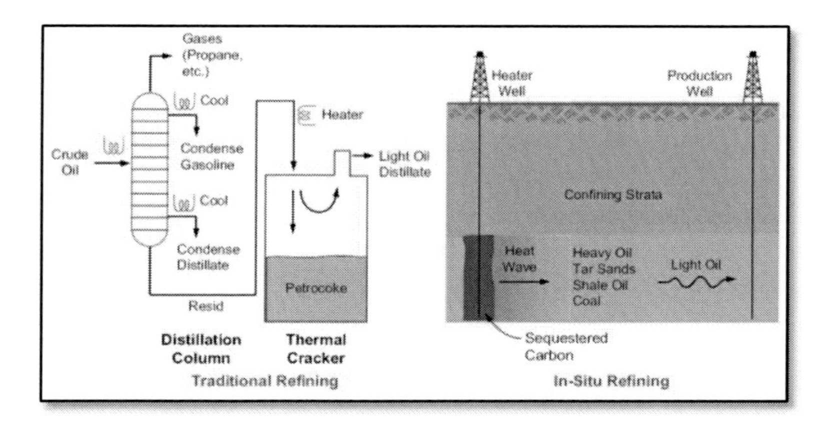

Figure 21. Distillation and thermal cracking of high-molecular-weight hydrocarbons in a refinery and in an underground reservoir (Forsberg, 2009).

This distillation process leaves most impurities behind; as the temperature further increases, heavier hydrocarbons that have not been vaporized will be thermally cracked and turn into lighter volatile hydrocarbons, that can be recovered. This process has two major technical advantages:

- Ability to extract deep-situated resources: approximately 80% of the oil shale deposits worldwide are too deep for surface mining and can only be recovered with in-situ methods (Finan, Miu, and Kadak, 2005);
- Control of carbon dioxide emissions: unlike in traditional refining, the solids from an underground thermal-cracking process remain sequestered underground as carbon (Forsberg, 2009); if the heat was provided by an energy source that did not emit carbon dioxide as well, such as nuclear heat the result would be low emissions of carbon dioxide from the entire process, since a high-quality crude oil is distilled that requires little added refining to produce transport fuels (Forsberg, 2008).

Nuclear Energy as a Source of Heat for In-Situ Shale Oil Recovery

The shale oil heating process requires large quantities of high-temperature heat - about one-sixth the heating value of the product (IAEA, 1997). The heating of oil shale yields both liquids and gases (Forsberg, 2009); it is currently proposed to burn the gases, representing one-third of the recovered energy, to produce electricity, that is in turn converted into heat for further underground heating (Forsberg, 2009).

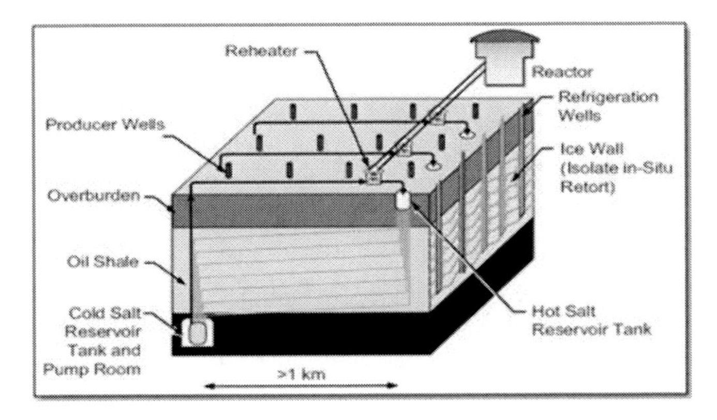

Figure 22. Configuration for underground heating of oil shale via nuclear heat (Forsberg, 2009).

This solution implies the release of greenhouse gases produced during the gas combustion. Although, a heating option exists that can maintain the process a greenhouse-free one: the use of high-temperature nuclear reactors to produce the required heat (see figure 22) (IAEA, 1997). Heat from nuclear plants guarantees two main advantages when compared to heat from the combustion of oil shale gases:

- It erases the necessity to burn part of the products to generate heat, substituting it with bleeded heat; the thermal energy cost in thus substantially decreased.
- It avoids emissions of carbon dioxide throughout the production process.

(Forsberg, 2006) identifies nuclear heat as a potentially viable thermal source because of a particular characteristic of many U.S. oil shale deposits: they are more than 200m thick and can yield up to 625 million barrels of oil per km^2. This means the concentrated layout of American shale-oil deposits make it practical and economically viable to transfer heat over limited distances from a reactor to the deposit.

Economics

There is uncertainty about the commercial viability of shale oil as a crude oil substitute: the 2005 study *Oil Shale Development in the United States: Prospects and Policy Issues*, (RAND, 2005), indicates that oil production based on in-situ refining can be profitable if crude oil prices consistently stay above at least $50 per barrel; the current price of crude oil is about 80 dollars per barrel, but the time horizon for the commercial development of in-situ technologies is more than 20 years (RAND, 2005); the oil price forecast is not reliable on such a long term, due to uncertainties in the development of crude oil consumption, extraction technologies, oil substitutes, and to the political instability of supplier countries.

Nuclear heat can make in-situ refining more economically competitive: the state-of-the-art Shell in-situ retorting process uses electric power as the source for down-hole heating; about 250 to 300 kilowatt-hours are required for down-hole heating per barrel of extracted product (RAND, 2005). Assuming electricity at $0.05 per kilowatt-hour, power costs for heating using electrically-generated heat amount to between $12 and $15 per barrel (crude oil equivalent).

Assuming nuclear power as a cost-zero source of heat, in-situ refining via nuclear heat could become competitive with crude oil prices above 35 to 38 dollars per barrel, which is less than half the current market price.

Of course, the commercial development of in-situ technology will require high investments: Shell reports that it has spent tens of million dollars in developing its in-situ conversion technology, and that a pre-commercial demonstration plant that would produce about 1,000 barrels per day will cost additional 200 million dollars (RAND, 2005). Further investments would be needed to reach a mature, commercially viable technology.

Environmental Considerations

If the economic feasibility of shale oil production is verified, there are issues that need to be reckoned on the environmental front, including (EASAC, 2007):

- Land use: large tracts of public land would need to handed over to the production and processing of oil shale. There would be the concomitant requirement of infrastructures: roads, power supply and distribution systems, pipelines, water storage and supply facilities; In-situ retorting would be less disruptive to the landscape than open-pit mining, nonetheless it would involve the drilling of a large number of wells. Due to the poor flow conditions within the shale, the wells would have to be drilled close to each other; the wells would need to be connected to an shale oil and gas treatment plant by a network of pipelines (RAND, 2005);
- Water quality: potential sources of water pollution include mine drainage, point-source discharges from surface operations associated with solids handling, retorting, upgrading, and plant utilities; there is little understanding of the long-term impact of the underground liquefaction

and gasification on groundwater quality, but it is envisaged to be a very disruptive one;

- Water consumption: estimated water requirements for mining and retorting range from 2.1 to 5.2 barrels of water per barrel of shale oil product; in-situ processing eliminates or reduces a number of these water requirements, but it would still require a considerable use of water for oil and gas extraction, post-extraction cooling, and products upgrading and refining.

Table 18. Main conclusions about possible applications

Dimension	Conclusion
Temperature required	Near term applications requiring low temperatures can be realized by commercial nuclear reactors, such as LWR and PHWR.
	Applications at medium temperature could be realized with gas reactors like AGR. However, the chapter will not focus on them for the following reasons: They are generation II reactors and it is unlikely to invest on them; There is no experience (and literature) in heat applications from them.
	Long term applications at high and very high temperature are highly innovative and constitute a major field of study: the commercialization of HTGR is envisaged for 2030.
Reactor size	Near term applications at low temperature can be realized using both large and small reactors. In particular, large reactors of generation III and III+ are available, while the commercialization of small, innovative reactors is envisaged for about 2016. There is a lot of experience and literature about nuclear heat applications at low temperature using traditional reactors.
	Long term applications, requiring high and very high temperature, are mainly addressed to small reactors such as Generation IV VHTR (2030).
EPZ size	Only district heating is strongly influenced by the EPZ size, as it requires a high density of people relatively near the reactor. If a large reactor is used, the NPP will be probably located far from the population centre which harnesses the heat, due to EPZ constraints: in this case, a way to exploit the unused area around the plant must be found. If a small, innovative reactor with enhanced safety is used, the EPZ could be reduced or even eliminated: however, this is not a certainty.
	All other applications are not highly influenced by EPZ as they do not require a large amount of people near the plant: on the contrary, it makes sense to assume that they can be located inside the EPZ, avoiding long and expensive heat transmission lines, and allowing the exploitation of the unused area around the plant.
	In order to reduce/eliminate EPZ, the option to locate the NPP offshore is very interesting.

CONCLUSION

Three main dimensions affect the development of the applications described in this second section:

1. Temperature required: low, medium or high;
2. Reactor size: traditional, large ones or small innovative ones;
3. EPZ size: small or large.

The main conclusions about possible applications, in relationship with these aspects, are discussed in Table 18.

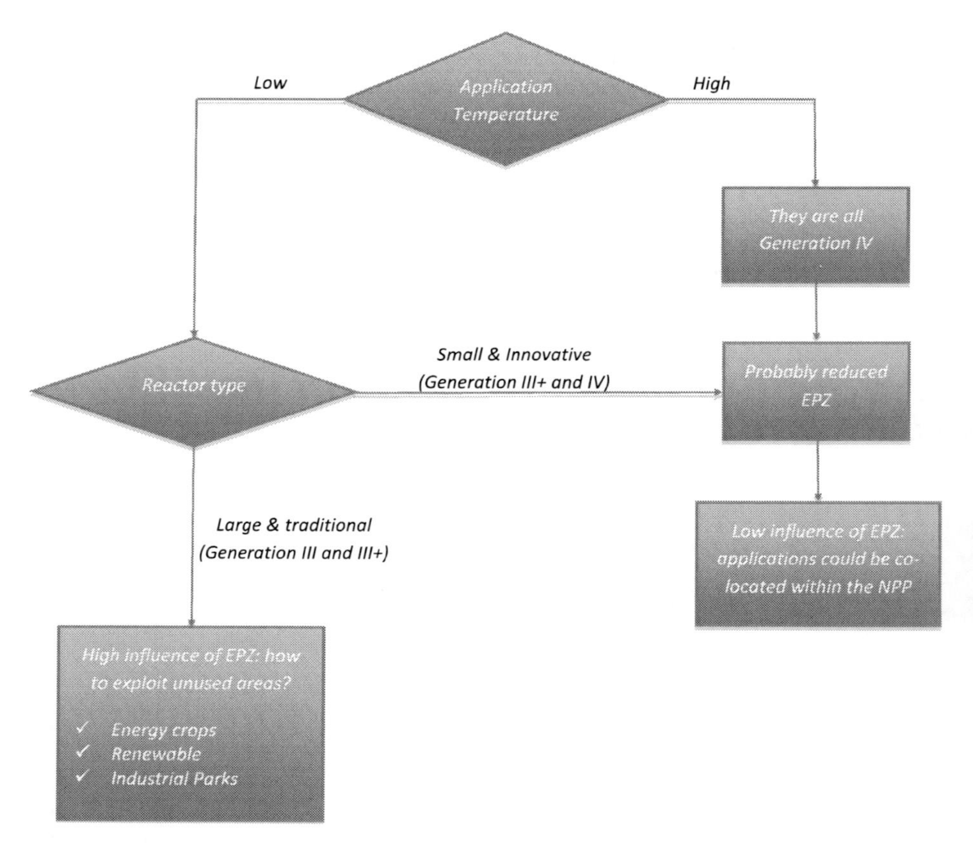

Figure 29. Influence of EPZ on the applications.

The hypothesized influence of the EPZ on the different applications can be schematized as follows:

- If the application considered is at low temperature and the reactor is a large, traditional one, the EPZ is hypothesized to be large and must be exploited in some way.
- If the application temperature is low, but the reactor is a small, innovative one (e.g. IRIS), the EPZ will be probably reduced or even collapsed in the NPP on-site area. Thus, the influence of EPZ in this case is not a major constraint;
- The same consideration apply for high temperature applications: they all envisage Generation IV reactors, for which the reduction of EPZ is a goal announced by GIF and INPRO.

These observations are schematized in Figure 29.

However it is not possible to assume the EPZ reduction for small reactors as a certainty. Even though the EPZ for these reactors is likely to be small, the very last decision is under responsibility of each country's legislation.

REFERENCES

AEC. (2005). Retrieved 2010 from Atomic Energy Council: http://www.aec.gov.tw/www/english/nuclear/psar/lm_document_psar_02_01.pdf

Augutis. (2005). Economic benefits of revising the need for relocation and evacuation measures unique to NPPs for innovative SMRs for regions with electricity co-generation. LEI.

BARC. (1975). *BARC Highlights: Reactor technologies and engineering*. India: BARC.

Blanchette. (2007). A hydrogen economy and its impact on the world as we know it. *Energy policy* .

Ciferno, and Marano. (2002). *Benchmarking biomass gasification technologies for fuels, chemicals and hydrogen production*. U.S. Department of Energy.

Conte, Iacobazzi, Ronchetti, and Vellone. (2001). Hydrogen economy for a sustainable development: state-of-the-art and technological perspectives. *Journal of Power Sources* .

Csik, K. (1997). *Nuclear power applications: supplying heat for hoime and industries*. IAEA.

EASAC. (2007). *A study on the EU oil shale industry*. EASAC.

Finan, Miu, and Kadak. (2005). *Nuclear technology and Canadian oil sands: integration of nuclear power with in-situ oil extraction*. Cambridge: Department of nuclear science and engineering.

Forsberg. (2006). High-temperature nuclear reactors for in-situ recovery of oil from oil shale. *Proceedings* .

Forsberg. (2008). High-temperature reactors for underground liquid-fuels production with direct carbon sequestration. *Proceedings* .

Forsberg. (2003). Hydrogen, nuclear energy, and the advanced high-temperature reactor. *International journal of hydrogen energy* .

Forsberg. (2009). Sustainability by combining nuclear, fossil, and renewable energy sources. *Progress in nuclear energy* .

Forsberg, and Peddicord. (2001). *Hydrogen production as a major nuclear energy application*. U.S. Department of Energy.

Hedges. (2005). A Brief Introduction to CANDU9 – A New Generation of Pressurized Heavy Water Reactors. From Canteach: http://canteach.candu.org/library/20054416.pdf

IAEA. (2007). *Advanced applications of water cooled nuclear power plants*. Vienna.

IAEA. (1997). *Floating nuclear energy plants for seawater desalination*. Vienna.

IAEA. (2000). IAEA TECDOC 1184 - *Status of non-electric nuclear heat applications: technology and safety*. IAEA.

IAEA. (2006). IAEA TECDOC 1487 - *Advanced nuclear plant design options to cope with external events*. IAEA.

IAEA. (2007). IAEA TECDOC 1584 - *Advanced applications of water cooled nuclear power plants*. IAEA.

IAEA. (1997). IAEA TECDOC 953 - *Method for the development of emergency response preparedness for nuclear or radiological accidents*.

IAEA. (2002). *Market potential for non-electric applications of nuclear energy*. IAEA.

IAEA. (2005). Meeting report of the first research coordination meeting (RCM) of coordinated research project (CRP) on small reactors without on-site refuelling. IAEA.

IAEA. (2003). *Method for Developing Arrangements for Response to a Nuclear or Radiological Emergency* (Updating IAEA TECDOC 953). IAEA.

IAEA. (1998). *Nuclear heat applications - Design aspects and operating experience*.

Kirchsteiger. (2006). Current practices for risk zoning around NPPs in comparison to other industry sectors. *Journal of Hazardous Materials*.

Komabe, Hanaoka, and Fujimoto. (2007). Co-gasification of woody biomass and coal with air and steam. *Fuel* .

Kutznetsov. (2008). The Technical and Economic Principles and Lines of Development of Nuclear District Heating Cogeneration. *Thermal engineering* , 55 (11), 926-938.

Levin, and Chahine. (2009). Challenges for renewable hydrogen production from biomass. *Hydrogen energy* .

Maioli. (2006). *EPZ History and background.*

McKendry. (2002). Energy production from biomass (part 3): gasification technologies. *Bioresource Technology* .

Minchener. (2005). Coal gasification for advanced power generation. *Science Direct* .

Mozaffarian, Zwart, Boerrigter, and Deurwaarder. (2004). *Biomass and waste-related SNG production technologies*. Rome.

NRC. (1998). NUREG 0654 - *Criteria fro preparation and evaluation of radiological emergency response plans and preparedness in supporto of NPP*. NRC.

NRC. (1988). NUREG 1140 - *A regulatory analysis on emergency preparedness for fuel cycle and other radioactive material licences*. NRC.

NRC. (1990). NUREG 1150 - Severe accident risk: an assessment for five US nuclear power plants. NRC.

NRC. (1998). PEP. From Nuclear Regulatory Commission: http://www.nrc.gov /about-nrc/emerg-preparedness/images/10-mile-EPZ-map.jpg)

NRC. (1998). *Regulatory guide 4.7 - General site suitability criteria for NPS.* Retrieved 2010 from U.S. NRC: http://www.nrc.gov/reading-rm/doc-collections/reg-guides/environmental-siting/active/04-007/

NRC. (1962). *TID-14844 Calculation of Distance Factors for Power and Test Reactor Sites*. NRC.

NRC. (2003). Title 10 - *Energy*, Volume 2, Part 100 - Reactor Site Criteria. NRC.

NRC. (1950). WASH-3. NRC.

OECD/NEA. (2004). *Non-electricity products of nuclear energy*. OECD.

OECD/NEA. (2003). *Short-term Countermeasures in Case of a Nuclear or Radiological Emergency.*

OPG. (2009). Site boundary consideration for new nuclear - Darlington. From Canadian Environmental Assesment Agency: http://www.ceaa.gc.ca/ 050/documents_staticpost/cearref_29525/0105/ai-sb.pdf

Ots. (2007). *Estonian oil shale properties and utilization in power plants.* Energetika .

RAND. (2005). *Oil shale development in the United States: prospects and policy issues*. U.S. Department of Energy.

RENAEL. (2004). *Vademecum sulle tecnologie del risparmio energetico: teleriscaldamento*. From http://www.renael.net/public/documenti/181/ Teleris caldamento.pdf

Ro, Cantrell, Elliott, and Hunt. (2007). *Catalytic Wet Gasification of Municipal and Animal Wastes*. From ACS Publications: http://pubs.acs.org

Schultz, Brown, Besenbruch, and Hamilton. (2003). *Large-scale production of hydrogen by nuclear energy for the hydrogen economy*. San Diego: General Atomics.

Stiegel, and Maxwell. (2001). Gasification technologies: the path to clean, affordable energy in the 21st century. *Fuel processing technology* .

Turn. (1999). Biomass integrated gasifier combined cycle technology III: application in the sugar cane industry. *Int. Sugar Journal* .

WNA. (2009). Generation IV nuclear reactors. From WNA: http://www.world-nuclear.org

WNA. (2009). *Nuclear power in Italy*. From WNA: http://www-world-nuclear.org

WNA. (2010 йил 12-March). *Nuclear power in Russia*. From WNA: www.world-nuclear.org

World Energy Council. (2007). 2007 *survey of energy resources*. London: World Energy Council.

In: Energy Resources
Editor: Enner Herenio de Alcantara

ISBN: 978-1-61324-520-0

Chapter 4

LIGNIN: A PEBBLE IN THE SHOE OF BIOETHANOL PRODUCTION

Iker Hernández *

Departament de Biologia Vegetal; Facultat de Biologia,
Universitat de Barcelona, Barcelona, Spain

ABSTRACT

Lignin is a complex polymer made of aromatic phenylpropanoid monomers such as sinapoyl, coumaroyl and coniferyl alcohols. Lignin deposits in plant cell walls -thus it is particularly abundant in heartwood-after polysaccharides, providing plant cell walls with physico-chemical properties key for water transport through xylem, plant defence and architecture. One of such properties is chemical resistance -recalcitrance-since lignin is extremely difficult to degrade either chemically or enzymatically, so lignin protects plant cell wall polysaccharides from degradation. The production of bioethanol from plant material lays on the de-polymerization of plant cell wall polysaccharides (saccharification) such as cellulose, followed by the alcoholic fermentation of the resulting reducing sugars. The presence of lignin in lingo-cellulosic feedstocks turns them recalcitrant to chemical attacks, and physically prevents the accession of cellulases and hemicellulases to their substrates during enzymatic saccharification. Lignin is the second most abundant biopolymer in terrestrial ecosystems (after cellulose), so its removal poses a major bottleneck for the production of bioethanol from lignocellulosic feedstocks.

---

* E-mail: iker_hernandez@ub.edu; Tel. +34 934021463; Fax. +34 934112842.

Nowadays, for this purpose, lignin is degraded by applying different pre-treatmets (de-lignification) such as extremely high temperature or pressure, strong acid/alkali attacks, etc. These pre-treatments reduce the net energy yield of the whole process, aside of generating abundant and hazardous residues and consuming time and resources. Thus, it is of capital importance to overcome the limitations that lignin imposes in the saccharification of lingo-cellulosic feedstocks if we are to produce bioethanol from them in an efficient manner. Research efforts are being invested in engineering plant cell walls to produce a less recalcitrant lingo-cellulosic material without imposing growth penalties to engineered plants.

In the presente manuscript the composition of plant cell wall polysaccharides, and the advances on lignin engineering for a more efficient production of plant material-based bioethanol production, are reviewed.

1. Plant Cell Wall Structure

The plant cell wall (cell wall herein) is an extra-cellular matrix immediately close to the plasma membrane the components of which are secreted by the plant cell. The cell wall is rigid and attached to the cell wall of the neighbouring cell(s), providing plant cells and tissues with mechanical features that are crucial for life (Reiter 2002). For instance xylem, the most important vascular tissue for water transport in vascular plants (over 90 % of the water transpired in leaves is transported through xylem), is essentially made of the remaining modified plant cell walls of dead cells. The mechanical properties of xylem cells (tracheids and vessel elements) allow this tissue to stand the strong negative pressures that drive water movement (Nobel 1999, Vermerris et al 2010).

Cell walls can be dissected in three main layers according to their structure, composition and properties: middle lamella, primary and secondary cell walls (Figure 1).

Middle lamella is the first cell wall layer synthesisized after cytokinesis (thus it is always the outermost layer) and derives directly from the cell division plate. This cell wall layer is found between adjacent cells and glues neighbouring cells between each other (Figure 1). Middle lamellae are mainly made of pectins and a small amount of structural proteins. It is often difficult to distinguish the limit between middle lamella and the primary cell wall.

The later, often called the growing cell wall, is made of crystalline cellulose microfibrills embedded in a matrix of hemicelluloses and pectins (the later two often called matrix polysaccharides). The primary cell wall is a stiff, yet dynamic structure that allows cell expansion upon loosening. Together with middle

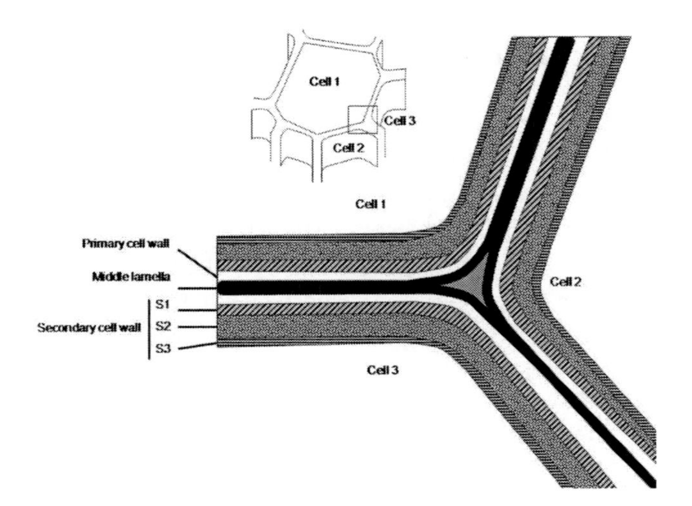

Figure 1. Schematic view of the plant cell wall including middle lamella, primary, and secondary cell walls. S1, S2 and S3 are sub-layers of the secondary cell wall in chronological order of deposition (thus, S3 is the innermost cell wall layer). The fill of the S1, S2 and S3 layers in the figure represent the alternating oritenation of cellulose microfibrills.

lamella, the primary cell wall accounts for most of the apoplast space and host the metabolites (structural proteins and enzymes, sugars, small organic acids, etc.) involved in apoplast metabolism such as cell wall dynamics, apoplast-symplast signalling, pathogen-host interaction, etc. (Cosgrove 2005, Pignocchi et al 2006).

Once a cell with primary cell wall has achieved its final size and shape, the secondary cell wall is deposited (thus, this is always the innermost cell wall layer). This cell wall layer consists on several sub-layers of cellulose microfibrils (named S1, S2, S3, etc. in order of deposition) oriented in fixed, normally alternating high and low, angles cemented by hemicelluloses and lignin (Figure 1).

The proportion the pectin fraction is far less abundant in secondary compared to primary cell walls. Moreover, the cellulose amount deposited in secondary cell walls is much higher than that in primary cell walls. Structural proteins can be also present in secondary cell walls, although in very small amounts (Ringli et al 2001, Seifert and Roberts 2007). In summary, aside of the molecules involved in apoplast metabolism (enzymes, small organic acids, etc.) and small amounts of structural proteins, cell walls are made of 4 quantitatively major structural polymers: pectins, hemicelluloses, celluloses and lignin, in variable proportions (Table 1).

2. Plant Cell Wall Composition

Pectins are galacturonic acid (GalA)-containing polysaccharides soluble in water and dilute acid solutions or calcium chelators (Cosgrove 2005). The abundance of pectins in cell walls varies strongly depending on the cell wall deposition stage, from ca. 30 % in some plants with no secondary cell wall, to less than 0,1 % in secondary cell walls of some species (Caffall and Mohnen 2009, Ishii 1997, Mellerowicz et al 2001). It has been estimated that about 90 % of the uronic acids in plant cell walls correspond to Gal*p*A (*p* stands for pyranosyl). Pectins, included amongst the most complex polymers in living organisms, are classified according to the nature the sugars present in the pectic polysaccharide aside of Gal*p*A. Homogalacturonans (HG) are the most simple and abundant pectins (depending on the species, over 60 % of the pectic dry weight). HGs are linear polymers of α-1,4-linked D-Gal*p*A that may be methylated or acetylated in the C6 or O2 positions, resp. Class I rhamnogalacturonans (RG-I) consist of a backbone of repeating units of Gal*p*A-rhamnose (Rha) $[\rightarrow \alpha$ -D-Ga*lp*A-1,2- α -L-Rha*p*-1,4$\rightarrow]_n$, branched to different extents at the C4 position of the Rha*p* residues with arabinan, galactan, arabinogalactan and some other minor side chains. Class II rhamnogalacturonans (RG-II) consist on a HG backbone of 7-9 Gal*p*A units with 4 well-defined side chains made of 12 different sugars, including some rare sugar species. RG-II tends to form dimers by boron diester bonds. Xylogalacturonans and apiogalacturonans are pectins found in aquatic plants (*Lemnaceae* and *Zosteraceae*, respectively) that contain xylose and apiose (or oligomers of these monosaccharides) respectively, covalently bound to the C3 position of Gal*p*A residues of an HG backbone. The diversity of side chains, the length of the HG or Gal*p*A-Rha*p* backbones, and the extent of methylation and acetylation of the Gal*p*A residues allow great chemical diversity and complexity in pectins (Burton et al 2010). Moreover, Ca^{2+} cations are able to ionically bind two different pectin chains by the "free" (i.e. non esterified) acid groups of the Gal*p* residues, so in the presence of this cation pectins appear as a hydrated jelly matrix. Both RG-I and RG-II are covalently bound to HG (and possibly to hemicelluloses), although the nature of the linker remains elusive (Caffall and Mohnen 2009). Thus, it may be thought of pectin matter as a continuous network with HG, RG-I and RG-II domains given cohesion by ionic bonds with cations such as Ca^{2+} and B^{3+}, rather than separate pectins bound to each other by ionic bonds. The chemical and structural diversity, along with the presence of cations, modulate cell wall functions such as adhesion, strengthen and loosening, stomatal movements and defence against pathogens.

Hemicelluloses represent about 30 % of plant biomass (Table 1). Hemicelluloses are polymers of β-(1,4)-linked pyranosyl residues with the O4 in the equatorial position, that can be extracted from the cell wall by alkaline solvents (for review, see Scheller and Ulvskov 2010). Hemicelluloses are present in both primary and secondary cell walls, although more abundant in the later. Moreover, monocots have more hemicellulose and less pectin in their primary cell walls compared to dicots (Table 1). The structure of hemicelluloses resembles that of celluloses, but branching of their linear structure avoids packing and crystallization of hemicelluloses in fibrils as it occurs for cellulose. Xylose is always the most abundant sugar in hemicelluloses. Several types of hemicelluloses can be distinguished, being xylans, xyloglucans and mannans the most abundant ones. Xylans are particularly abundant in monocots, and consist of linear chains of β-(1,4)-linked D-xylose (Xyl) residues (homoxylans) that may be branched with glucuronic acid (glucuronoxylans), arabinose (arabinoxylans), both (glucuronoarabinoxylans), and/or more rarely 4-*O*-methylglucuronic acid, in the O2 and O3 positions of the homoxylan units. Xyloglucans are the most abundant hemicellulose type in dicots. Their structure is β-(1,4)-linked D-glucose (Glc) (like cellulose) abundantly substituted at the C6 position by xylose residues. In additiobn xyloglucans may be also substituted with galactose, fucose or arabinose, and acetylated. Hemiceluloses bind cellulose microfibrills (see below) tightly by abundant hydrogen bonds. Early binding of hemicelluloses in the primary wall to growing cellulose microfibrills results in thinner cellulose microfibrils compared to those of secondary cell walls, which renders primary cell walls more flexible. The binding of hemicelluloses to cellulose microfibrills provides cellulose networks with an order due to the repeating structure of the formers (backbone and side chains), which varies depending on the hemicellullose properties and the binding pattern. Mannans are a minor type of hemicelluloses that aside of the structural function, are thought to act as energy reservoirs (Reid 1985). Moreover, arabinogalactans are also considered hemicelluloses although they are water-soluble in the presence of EDTA, like pectins. Arabinogalactans are typically bound to proteins (arabinogalactan proteins, AGPs), some of the most important proteins (quantitatively) in plant cell walls, together with prolin-rich proteins, glycine-rich proteins and wall-associated protein kinases.

Cellulose represents about 50 % of plant biomass and is the most abundant biopolymer in terrestrial ecosystems. Cellulose consists of large linear chains of 2000-25000 β-(1,4)-linked D-glucose (Glc) monomers that can wind several times a single cell (1-5 μm). Cellulose is organized in highly crystalline, yet with amorphous regions, microfibrills that consist on parallel cellulose chains tightly packed by hydrogen bonds and Van der Waals forces. Moreover, as stated before, hemicelluloses bind celluloses either by surface interactions or by becoming physically trapped between several tightly attached glucan chains of the microfibril.

The structure of the cellulose synthase macromolecular complex, which is attached to the plasma membrane, determines the shape of cellulose microfibrils: a single globular unit of cellulose synthase A (CesA) synthetizes a glucan chain; typically 6-10 globular units of CesA form a rosette subunit that produces the so-called 2-nm fibers (fibers of 6-10 glucan chains), and 6 rosette subunits cluster together to form a rosette (a structure bound to the plasmalema) that yields cellulose microfibrills of typically 36 glucan chains (ca. 30 nm diameter microfibrils).

The insoluble and highly crystallized nature of the cellulose microfibrils provide them with high tensile strength, and resistance to enzymatic or chemical attacks. Still, some microorganisms and some chemical and physical treatments are able to degrade cellulose yielding mono and oligo-saccharides that are excellent substrates for ethanolic fermentation, altogether making cellulose the main target of the saccharification of lingo-cellulosic feedstocks (Sims et al 2010).

Finally, lignin adds an extra degree of complexity to plant cell walls. It is a polymer that branches in all three dimensions and binds covalently to hemicelluloses. It is mainly deposited in secondary cell walls, although it may be present in tiny amounts in primary cell walls, representing about 30 % of the organic carbon in the biosphere (Boerjan et al 2003). Lignin contributes to strengthen cell walls and provides a hydrophobic environment necessary for water transport through xylem.

Table 1. Typical composition of plant cell walls from different feedstocks. Values express percentage of the dry cell wall weight. [1]Poales are taken as paradigmatic monocots, although other monocots show more dicot-like cell wall composition ([a]Ververis et al 2007; [b]Domozych et al 1980; [c]Galbe and Zacchi 2002; [d]Calfall and Mohnen 2009; [e]Uzal et al 2009; [f]Fry 1988; [g]Ishii 1997; [h]Ebringerová et al 2005; [i]Vogel 2008)

	Green algae	Gymnosperms	Grasses[1]	Dicots
Cellulose	7-17[a]	43-45[c]	25-35[f,g]	45-50[f,g]
Hemicelluloses	16[a]	20-23[c]	40-50[h]	25[h]
Pectins	30-39[b]	30[d]	0,1[g]	0,1[g]
Structural proteins	22-37[b]	Traces[e]	0-1[i]	0-1[i]
Lignin	Absent[b]	28[c]	7-10[g]	20[g]

Due to the complexity and variety of bonds between different monolignols (lignin subunits), and the stated hydrophobicity, lignin is it is extremely difficult to degrade in biological conditions, either by enzymatic or chemical attacks.

It is estimated that 90 % of the world-wide biomass production, estimated to be about $200x10^9$ Tm per year, is made of lingo-cellulosic material (i.e. plant cell walls) (Amthor and Baldocchi 2001). For instance, wood is a particularly rich source of this material, and is made of about 50 % cellulose, 25 % lignin, 25 % hemicelluloses and traces of pectins and structural proteins.

3. Biosynthesis of Lignin

3.1. Biosynthesis of Monolignols

As introduced before, lignin is a highly branched polymer of phenylpropanoid units bound to each other by oxidative coupling. Lignin subunits derive form the aromatic aminoacid phenylalanine (and in minor amounts, form tyrosine) that is synthetized by the shikimate pathway (for review, see Herrmann and Weaver 1999). By the sequential action of PAL (phenylalanine ammonia lyase), C4H (cinnamate 4-hydroxylase) and 4CL (4-coumarate:CoA ligase), phenylalanine is transformed into *p*-coumaroyl-CoA: the precursor the three most abundant lignin building blocks (monolignols): *p*-coumaryl, coniferyl, and sinapyl alcohols. For the formation of *p*-coumaryl alcohol, *p*-coumaroyl-CoA is sequentially reduced to *p*-coumaraldehyde and *p*-coumaryl alcohol by the action of CCR (cinnamoyl-CoA reductase) and CAD (cinnamyl alcohol dehydrogenase), resp. For the synthesis of coniferyl and sinapyl alcohols, *p*-coumaroyl-CoA is sequentially converted in coniferaldehyde, a common precursor of both coniferyl and synapil alcohols, by the action of HCT (*p*-hydroxycinnamoyl-CoA:quinate shikimate *p*-hydroxycinnamoyltransferase), C3H (*p*-coumarate 3-hydroxylase), HCT, CCoAMT (caffeoyl-CoA *O*-methyltransferase), and CCR (cinnamoyl-CoA reductase). Coniferaldehyde may be reduced to coniferyl alcohol by the action of CAD, or further modified by F5H (ferulate 5-hydroxylase) and COMT (caffeic acid/5-hydroxyconiferaldehyde *O*-methyltransferase) to sinapaldehyde, which is further reduced to sinapyl alcohol by SAD (sinapyl alcohol dehydrogenase). In addition to this main monolignol biosynthetic pathways, a complex array of reactions connect the different branches of monolignols biosynthesis, creating a biosynthetic network with a wide range of different phenylpropanoid intermediaries and end-products. Although *p*-coumaryl, coniferyl and sinapyl alcohols are the "classic" (and most abundant) lignin building blocks it has become clear that other phenylpropanoids from this biosynthetic grid, such as caffeyl alcohol and ferulic acid, are part of the lignin polymer. These additional units are basically intermediaries of the biosynthesis or derivatives of the major monolignols (Figure 2).

It is noteworthy that monolignols are synthetized in the cytosol (probably anchored to the cytosolic face of the endoplasmic reticulum), so they need to be transported to the cell wall. The mechanisms for the transport of monoliognols are still unclear, but the involvement of vesicle trafficking of glycosilated monolignols has been suggested (Vanholme et al 2008).

In most cases, the biosynthesis of lignin is co-regulated with that of other secondary cell wall components such as cellulose and hemicelluloses. Most lignin biosynthetic genes have common *cis*-acting elements in their promoter regions (AC-elements and G-boxes) that make them responsive to the same transcription factors. The integrated control of the biosynthesis of lignin and other secondary cell wall components is carried out by a complex network of NAC (SND1, NST1, VND6 and homologues) and MYB transcription factors (e.g. MYB26 and MYB 46) that bind the *cis*-elements cited above (Zhong and Ye 2007). On one side, NAC transcription factors act as master switches of the lignin biosynthetic genes. On the other hand, MYB transcription factors are master switches of the secondary cell wall biosynthetic genes (i.e. cellulose, hemicellulose and lignin biosynthetic genes) and targets of NAC transcription factors. Lignin biosynthesis and deposition is strongly affected by all kind of external (e.g. water deficit, excess light, UV-B radiation, mechanical stress, pathogen attacks, etc.) and internal (hormones, developmental signals, etc.) stimuli that regulate the expression of the NAC and MYB master regulators (Zhong and Ye 2007).

3.2. Lignification

p-Coumaroyl, coniferyl and sinapyl alcohols, which differ mainly in the degree of methoxylation, originate hydroxyphenyl (H), guaiacyl (G) and sinapyl (S) subunits in the lignin polymer after oxidative coupling. The proportion of H, G and S residues in lignin varies among species, tissues, developmental stages, etc, but in general, angiosperm dicots tend to have G and S with traces of H units; gymnosperms, G with traces of H units; and angiosperm monocots, G and S units at equivalent amounts, and H units at higher amounts than dicots (Boerjan et al 2003). Lignification occurs after dehydrogenation of monolignols to generate the radicals to be coupled. This dehydrogenation is carried out by different cell wall enzymes such as peroxidases, laccases and polyphenol oxidases, but it is still unclear how lignification is occurs *in vivo*. The classic view is that in which monolignols are oxidized and coupled in a combinatorial way. This implies that combinatorial monolignol radical-coupling is a simple chemical reaction in which

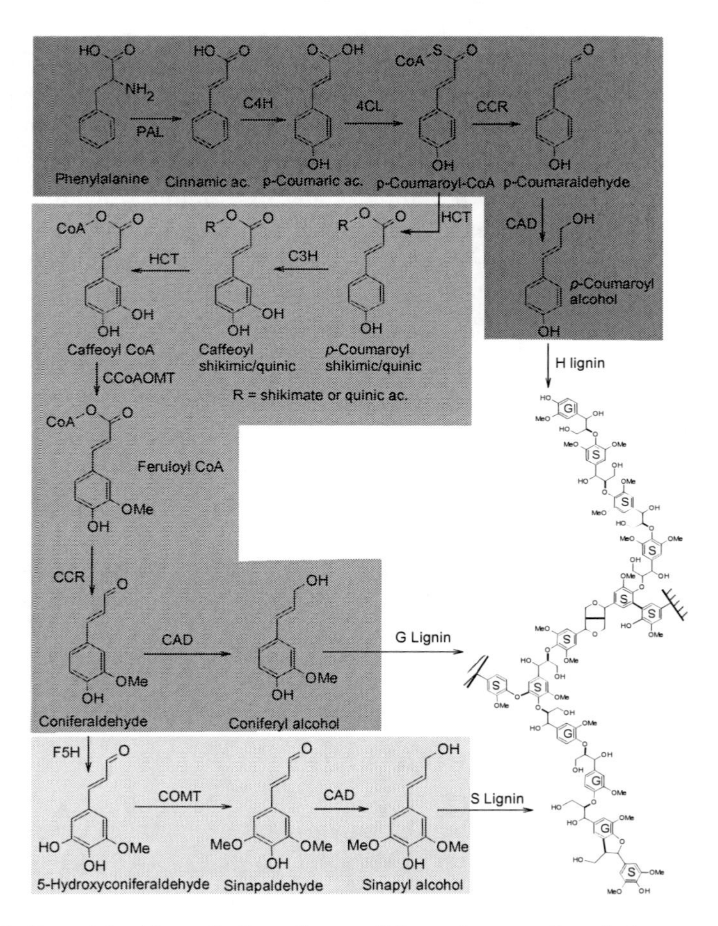

Figure 2. Outline of the biosynthetic pathway of the major monolignols (coumaryl, coniferyl and sinapyl alcohols), and a model angiosperm lignin molecule showing some of the most common bonds resulting from radical-coupling (modified from Vanholme et al 2008, and Boerjan 2003). PAL, phenylalanine ammonia lyase; C4H, cinnamate 4-hydroxylase; 4CL, 4-coumarate:CoA ligase; HCT, *p*-hydroxycinnamoyl-CoA:quinate shikimate *p*-hydroxycinnamoyl transferase; C3H, *p*-coumarate 3-hydroxylase; CCoAOMT, caffeoyl-CoA *O*-methyltransferase; CCR, cinnamoyl-CoA reductase; F5H, ferulate 5-hydroxylase; COMT, caffeic acid//5-hydroxycaniferaldehyde *O*-methyltransferase; CAD, cinnamyl alcohol dehydrogenase. H, S and G stand for hydroxyphenyl, guayacyl and syringyl lignin units.

any phenolic compound present in the lignification area could be involved in, if chemical conditions (e.g. pH, temperature, presence of H_2O_2 and peroxidases) allow it (Freudenberg 1968). Other authors, however, support the existence of dirigent proteins that drive the monolignol radical-coupling by the means of dirigent sites (Davin and Lewis 2005). This view suggests that monolignol composition and lignin structure are biochemically regulated, but such a regulation is still unclear (Vanholme et al 2008). Five main bonds can be observed in the lignin polymer: β-β (or resinol, binds the β carbon of two lignin precursors), β -*O*-4 (or β -ether; binds the β carbon of the incorporating monolignol to the growing polymer through an ether bond), β -5 (or phenylcoumaran; binds the β carbon of the incorporating monolignol to the phenolic ring of other monolignols), 5-5 (or biphenyl; binds the C5 of the incorporating monolignol to the C5 of other monolignols), 4-*O*-5 (biphenylether; binds the C5 of the incorporating monolignol to the C4 of an existing monolignol through an ether bond). The β-ether and phenylcoumaran units result from the incorporation of a single monolignol to an existing lignin molecule, while the biphenyl and biphenyl ether units normally arise from the coupling of two pre-existing oligomers (Boerjan et al 2003). The fifth type of bond (resinol) is formed exclusively when two monolignol precursors dimerize. In addition, different end-groups can be found, namely ferulic acid, cinnamyl alcohol, etc. (Figure 2). The β-ether bounds of lignin are the most abundant and least recalcitrant to chemical cleavage.

4. Interference of Lignin in the Saccharification Process

Cellulose is the most abundant biopolymer in the Earth. It is produced at a rate of ca. 10^{11} tons per year, and it shows a very high energetic content, so it is an extremely good potential substrate for the production of biofuels (Sims et al 2010). But as Mansfield put it, plant cell wall is "a wall built to last" (Mansfield 2009). If cellulose is the most abundant terrestrial biopolymer, lignin follows it representing about 30 % of the organic C (Boerjan et al 2003). As stated before, lignin binds hemicelluloses covalently, this way creating a recalcitrant waterproof shield that protects cellulose microfibrills from enzymatic and mild chemical attacks. For the generation of bioethanol from ligno-cellulosic feedstocks, polysaccharides need to be released from the protection of lignins (de-lignification) and saccharified to substrates fermentable by microorganisms. Nowadays, de-lignification is achieved by water-based acidic or alkaline, ammonium, or organic solvent pulping, steam explosion, and other chemical pre-treatments that generate hazardous and/or abundant waste, require considerable amounts of energy and degrade part of the sugars released. In addition, these treatments often generate by-products that inhibit the activity of the fermenting microorganisms, altogether strongly reducing the efficiency of the whole process (Galbe and Zacchi 2007). On the other hand, although more environmentally friendly, biological de-lignification is far less efficient. Engineering lignin metabolism (mainly composition and structure) has become a cornerstone for a more efficient de-lignification and, ultimately, a more efficient bioethanol production.

5. Lignin Engineering

Efforts for engineering monolignol biosynthesis have yielded reduced lignin amount, altered monolignol composition of lignin, appearance of novel monolignols in the lignin, or combinations of these phenomena. In general the down-regulation of monolignol biosynthetic genes (PAL, C4H, 4CL, HCT, C3H, CCoAOMT, CCR and CAD) reduces lignin amounts and changes monolignol proportions in it. Plants with down-regulated HCT and C3H are rich in H units, which are rare in wild type lignin. Downregulation of F3H results on lignin made essentially of G units, while downregulation of F5H results in lignin made essentially of S units. The downregulation of CAD leads to the incorporation of hydroxycinnamaldehyde, a "non-classic" monolignol, to lignin (reviewed by Vanholme et al 2008). The analysis of alfalfa (*Medicago truncatula*) lines with downregulated C4H, HCT, C3H, CCoAOMT, F5H or COMT, Arabidopsis (*Arabidopsis thaliana*) *4cl1* and *ccr1* KO lines, and poplar (*Populus trychocarpa*) CCR-downregulated plants show altered lignin composition and improved saccharification efficiency, while the concomitant shift in monolignol composition does not seem to impose a handicap for plant growth (Chen and Dixon 2007, Vermerris et al 2007, Vanholme et al 2010).

Still, since lignin fulfils key roles in terrestrial plant growth and development, extremely low lignin content often impairs plant growth and development. This impairment has been attributed to deficient water transport and mechanical support, but it has been shown that it can also be due to pleiotropic effects. For instance, Arabidopsis plants devoid of HTC deposit less lignin that is enriched in H units, which results on a reduction of growth; this growth reduction was shown to be due to the accumulation of flavonoids that inhibit auxin transport, and could be reverted by downregulating chalcone synthase, the enzyme that catalyzes the first committed step of flavonoid biosynthesis (Besseau et al 2007). The *cad-n1* allele is a naturally-occurring null allele of CAD in loblolly pines (*Pinus taeda*). Loblolly pines homozygous for this allele show reduced growth but their wood shows improved saccarification efficiency. Interestingly, heterozygous pines are higher and show enhanced wood density. The case of the *cad-n1* allele indicates that natural genotypic variation can provide novel traits for classic breeding strategies towards an efficient saccharification of lingo-cellulosic material (Yu et al 2006).

Other efforts have focused on manipulating lignin polymerization. Since the regulation of lignin polymerization is still obscure, these efforts have focused on the activation of monolignols for radical coupling i.e., the oxidation/dehydrogenation of monolignols. It has been proposed that peroxidases (PRXs) and polyphenol oxidases may be responsible for this oxidation. PRXs use H_2O_2 to oxidize their substrates, while polyphenol oxidases use O_2. Members of these two large enzyme families have been found in lignifying cell walls. However, tobacco plants with down-regulated PRX activity showed reduced lignin content, whereas downregulation of laccase (a polyphenol oxidase) activity in transgenic poplar plants did not affect lignin content. On the other hand, laccase 15 KO Arabidopsis plants (*lac15*) showed reduced lignin amounts in the seed coat. Altogether suggests peroxidases are key for monolignol oxidation, but that the process differs depending on the tissue, organ, species, etc. (Blee et al. 2003, Ranocha et al. 2002). It has been shown that the availability of H_2O_2 in the cell wall drives lignification in *Zinnia elegans* suspension cultures, but it is still to be demonstrated that this is the case in other plants (Fagerstedt et al 2010). H_2O_2 in cell walls may be produced by a number of enzymes (e.g. NADPH oxidases, copper amine oxidases, oxalate oxidases and peroxidases), but in Norway spruce (*Picea abies*) tissue cultures the cell wall-associated NADPH oxidase rboh (respiratory burst oxidase homolog) along with peroxidases seem to be the main sources of H_2O_2 (Kärkönen et al 2009). In addition, antioxidants also control H_2O_2 presence in the cell wall (Pignocchi and Foyer 2003). The levels of ascorbate, which is the main H_2O_2 scavenger in the plant cell wall, are strongly lowered in lignifying tissues, which provides indirect evidence that the manipulation of apoplastic ascorbate metabolism may be a potential target for lignin engineering. In summary attempts to reduce lignin amounts or increase lingo-cellulosic feedtstock saccharification potential through the manipulation of monolignol oxidation have yielded promising results, but the knowledge of this aspect of the lignification process is still scarce. As stated before, the transport of monolignols from the site of biosynthesis (the cytosol) to the lignification site (the apoplast) is still unclear. It is thought that coniferin and syringin (coniferyl alcohol and sinapyl alcohol 4-*O*-glucosides) are substrates of vesicle-mediated transport, given that the glucosilated monolignols are water-soluble. This would mean that monolignols are glycosilated in the biosynthetic site, transported, and de-glycosilated in the lignification site. However, downregulation of coniferyl and sinapyl alcohol 4-*O*-glucosyltransferases, while reduce the levels of coniferin and syringin, do not seem to affect lignin deposition (Vanholme et al 2008). These results raise the possibility of other monolignol transport mechanisms such as direct plasma membrane pumping by ABC or other type of transporters (Boerjan 2003). Thus, engineering monolignol transport is far from being a target for lignin modification, but the discovery of monolognol transport mechanisms would settle the basis for metabolic engineering of lignification.

6. Perspectives in Lignin Engineering

Research on lignin composition and polymerization has been boosted in the recent years due to the promising perspective of sustainable fuel production. However, we are far from producing ethanol from ligno-cellulosic feedstocks in an efficient manner. Understanding lignin metabolism is a cornerstone in bioethanol production, but certain aspects of lignin biochemistry are still obscure. On one side, the possibility of directed lignin polymerization should be explored further, since the possible factors determining lignification patterns would be major targets for lignin engineering. On the other hand, lignin is also a highly energetic polymer. Its recalcitrance imposes important limitations in the saccharification processs, but different physico-chemical treatments may be able to take advantage of this high energetic content (Munasinghe and Khanal 2010). Another obscure side of lignin metabolism is how monolignols are transported to the lignification site. Unravelling these mechanisms would offer new targets for lignin engineering. Finally, although radical coupling is the polymerization mechanism of lignin, it is not well understood the factors that drive the generation of monolignol radicals. Thus, research on the redox reactions (e.g. radical scavenging, enzymatic generation of free radicals, etc.) in the plant cell wall would provide us with more targets for lignin engineering.

REFERENCES

Amthor JS, Baldocchi DD. (2001). Terrestrial higher plant respiration and net primary production. In: *Terrestrial Global Productivity*, Academic Press, 33-59.

Besseau S, Hoffmann L, Geoffroy P, Lapierre C, Pollet B, Legrand M. (2007). Flavonoid accumulation in Arabidopsis repressed in lignin synthesis affects auxin transport and plant growth. *Plant Cell* 19:148-162.

Blee KA, Choi JW, O'Connell AP, Schuch W, Lewis NG, Bolwell GP. (2003). A lignin-specific peroxidase in tobacco whose antisense suppression leads to vascular tissue modification. *Phytochemistry* 64:163-176.

Boerjan W, Ralph J, Baucher M. (2003). Lignin biosynthesis. *Annu. Rev. Plant Biol.* 54:519-546.

Burton RA, Gidley MJ, Fincher GB. (2010). Heterogeneity in the chemistry, structure and function of plant cell walls. *Nature Chem. Biol.* 6:724-732.

Calfall KH, Mohnen D. (2009). The structure, function, and biosynthesis of plant cell wall pectic polysaccharides. *Carbohydr. Res.* 344:1879-1900.

Calfall KH, Mohnen D. (2009). The structure, function, and biosynthesis of plant cell wall pectic polysaccharides. *Carbohydr. Res.* 344:1879-1900.

Chen F, Dixon RA. (2007). Lignin modification improves fermentable sugar yields for biofuel production. *Nat. Biotechnol.* 5:759-761.

Cosgrove DJ. (2005). Growth of the plant cell wall. *Nat. Rev. Mol. Cell Biol.* 6:850-861.

Davin LB, Lewis NG. (2005). Lignin primary structures and dirigent sites. *Curr. Opin. Plant Biol.* 16:407-415.

Domozych DS, Stewart KD, Mattox KR. (1980). The comparative espects of cell wall chemistry in the gree algae (Chlorophyta). *J. Mol. Evol.* 15:1-12.

Ebringerová A, Ebringerová Z, Heinze T. (2005). Hemicellulose. *Adv. Polym. Sci.* 186:1-67.

Fagerstedt KV, Kukkola EM, Koistinen VV, Takahashi J, Marjamaa K. (2010). Cell wall lignin is polymerised by class III secretable plant peroxidases in Norway spruce. *J. Integr. Plant Biol.* 52:186-94.

Freudenberg K, Neish AC (eds.). (1968). *Constitution and biosynthesis of lignin.* Springer-Verlag, Berlin.

Fry SC (ed.). (1988). *The growing plant cell wall: chemical and metabolic analysis*. Longman Scientific and Technical, Essex, UK.

Galbe M, Zacchi G. (2002). A review of the production of ethanol from softwood. *Appl. Microbiol. Biotechnol.* 59:618-628.

Galbe M, Zacchi G. (2007). Pretreatment of lignocellulosic materials for efficient bioethanol production. *Adv Biochem. Eng. Biotechnol.* 108:41-65.

Herrmann KM, Weaver LM. (1999). The shikimate pathway. *Annu. Rev. Plant Physiol. Plant Mol. Biol. 50:*473-503.

Ishii T. (1997). Structure and functions of feruloylated polysaccharides. *Plant Science,* 127:111-127.

Kärkönen A, Warinowski T, Teeri TH, Simola LK, Fry SC. (2009). On the mechanism of apoplastic H_2O_2 production during lignin formation and elicitation in cultured spruce cells--peroxidases after elicitation. *Planta,* 2303:553-567.

Mansfield SD. (2009). Solutions for dissolution--engineering cell walls for deconstruction. *Curr. Opin. Biotechnol.* 20:286-294.

Mellerowicz EJ, Baucher M, Sundberg B, Boerjan W. (2001). Unraveling cell wall formation in the woody dicot stem. *Plant Mol. Biol.* 47:239-274.

Munasinghe PC, Khanal SK. (2010). Biomass-derived syngas fermentation into biofuels: Opportunities and challenges. *Biores. Tech.* 101:5013-5022.

Nobel PS (ed.). (1999). *Physicochemical and environmental plant physiology*, 2nd ed. Academic press, San Diego, CA.

Pignocchi C, Foyer CH. (2003). Apoplastic ascorbate metabolism and its role in the regulation of cell signalling. *Curr. Opin. Plant Biol.* 6:379-389.

Pignocchi C, Kiddle G, Hernández I, Foster SJ, Asensi A, Taybi T, Barnes J, Foyer CH. (2006). Ascorbate oxidase-dependent changes in the redox state of the apoplast modulate gene transcript accumulation leading to modified hormone signaling and orchestration of defense processes in tobacco. *Plant Physiol.* 141:423-435.

Ranocha P, Chabannes M, Chamayou S, Danoun S, Jauneau A, (2002). Laccase down-regulation causes alteration in phenolic metabolism and cell wall structure in poplar. *Plant Physiol.* 129:145-155.

Reid JSG. (1985). Cell wall storage carbohydrates in seeds—biochemistry of the seed "gums" and "hemicelluloses". *Adv. Bot. Res.* 11:125-155.

Reiter WD. (2002). Biosynthesis and properties of the plant cell wall. *Curr. Opin. Plant B*iol. 5:536-42.

Ringli C, Keller B, Ryser U. (2001). Glycine-rich proteins as structural components of plant cell walls.*Cell Mol Life Sci.* 58:1430-1441.

Scheller HV, Ulvskov P. Hemicelluloses. Annu. Rev. Plant Biol. 61:263-289.

Seifert GJ, Roberts K. (2007). The biology of arabinogalactan proteins. *Annu. Rev. Plant Biol.* 58:137-61.

Sims RE, Mabee W, Saddler JN, Taylor M. (2010). An overview of second generation biofuel technologies. *Bioresour. Technol.* 101:1570-1580.

Uzal EN, Gómez-Ros LV, Hernández JA, Pedreño MA, Cuello J, Ros Barceló A. (2009). Analysis of the soluble cell wall proteome of gymnosperms. *J. Plant Physiol.* 166:831-843.

Vanholme R, Morreel K, Ralph J, Boearjan W. (2008). Lignin engineering. *Curr. Opin. Plant Biol.* 11:278-285.

Vanholme R, Ralph J, Akiyama T, Lu F, Pazo JR, Kim H, Christensen JH, Van Reusel B, Storme V, De Rycke R, Rohde A, Morreel K, Boerjan W. (2010). Engineering traditional monolignols out of lignin by concomitant F5H1-up- and COMT-down-regulation in Arabidopsis. *Plant J.* (doi: 10.1111/j.1365-313X.2010.04353.x).

Vermerris W, Saballos A, Ejeta G, Mosier NS, Ladisch Mr, Carpita NC. (2007). Molecular breeding to enhance ethanol production from corn and sorghum stover. *Crop. Sci.* 47:S142-S153.

Vermerris W, Sherman DM, McIntyre LM. (2010). Phenotypic plasticity in cell walls of maize brown midrib mutants is limited by lignin composition. *J. Exp. Bot.* 61:2479-2490.

Ververis C, Georghiou K, Danielidis D, Hatzinikolaou DG, Santas P, Santas R, Corleti V. (2007). Cellulose, hemicelluloses, lignin and ash content of some organic materials and their suitability for use as paper pulp supplements. *Bioresource Technol.* 98:296-301.

Vogel J. (2008). Unique aspects of the grass cell wall. *Curr. Opin. Plant Biol.* 11:301-307.

Yu Q, Li B, Nelson CD, McKeand SE, Batista VB, Mullin TJ. Association of the cad-n1 allele with increased stem growth and wood density in full-sib families of loblolly pine. *Tree Grenet. Genomes* 2:98-108.

Zhong R, Ye ZH. (2007). Regulation of cell wall biosynthesis. *Curr. Opin. Plant Biol.* 10:564-572.

In: Energy Resources
Editor: Enner Herenio de Alcantara
ISBN: 978-1-61324-520-0

Chapter 5

CARBON DYNAMIC AND EMISSIONS IN BRAZILIAN HYDROPOWER RESERVOIRS

Jean P. Ometto[1], Felipe S. Pacheco[1,2], André C. P. Cimbleris[3], José L. Stech[1], João A. Lorenzzetti[1], Arcilan Assireu[1,7], Marco A. Santos[4], Bohdan Matvienko[5], Luiz P. Rosa[4], Corina Sidagis Galli[6], Donato S. Abe[6], José G. Tundisi[6], Nathan O. Barros[2], Raquel F. Mendonça[2] and Fabio Roland[2]

[1]Brazilian Institute for Space Research (INPE) São José dos Campos, Brazil
[2]Federal University of Juiz de Fora (UFJF) / Juiz de Fora, MG, Brazil
[3]FURNAS Centrais Elétricas / Rio de Janeiro, RJ, Brazil
[4]Federal University of Rio de Janeiro (COPPE/UFRJ), Rio de Janeiro, RJ, Brazil
[5]Construmaq Sâo Carlos, CP 717 / Sâo Carlos SP, Brazil
[6]Internation Institute of Ecology (IIEGA) / São Carlos, SP, Brazil
[7]Federal University of Itajuba (UNIFEI) / Itajubá, MG, Brazil

ABSTRACT

As well documented by the scientific literature and strongly stated by the Intergovernmental Panel on Climate Change (IPCC), changes in the atmosphere composition, due to increasing emissions of greenhouse gases

(GHG), may result in drastic environmental consequences, both in a regional and in a global scale. A substantial portion of the annual GHG emissions comes from the production of energy, which, in general, is produced from non-renewable sources. Coal dominates the electricity generation globally, accounting for 42 percent of total generation in 2007. As a renewable alternative, hydropower accounts to 18% of the global electricity production. However, concerns about the relation between carbon emission per unit of energy produced were raised in the scientific community in recent years. This concern is particularly important in tropical and sub-tropical regions, where temperature and the amount of flooded organic matter can be a fueled combination on the emission of greenhouse gases, in special methane. From 2003 through 2008 an extensive research program was carried out in 8 large reservoirs located in a broad geographical distribution in Brazil. The major goals on the project were: to determine the emissions of GHG to the atmosphere; to identify the pathways of the carbon cycle in the reservoirs; to evaluate the influence of morphometric, biogeochemical, biological and operational variables on the GHG emissions; and to develop a spatial and temporal model of the greenhouse gas emissions in reservoirs. In this chapter our aim is to report results and integrative outcomes from this study, as well as policy recommendation on managing hydropower systems in tropical regions.

Keywords: renewable energy, hydropower, carbon, greenhouse gases, climate change.

1. INTRODUCTION

The energy production to fuel the contemporary global economy has a massive participation of non-renewable sources, accounting for substantial portion of the greenhouse gases released to the atmosphere. According to Friedlingstein et al. (2010) and the Global Canopy Programme (www.globalcarbonproject.org) in 2009 a total of 8.4±0.5 PgC was emitted to the atmosphere by fossil fuel burning, being coal the largest portion of the emissions. This amount accounts to more than 85% of all anthropogenic carbon annually emitted as CO_2 to the atmosphere. With the emissions reduction observed in developed countries, the emissions numbers reflect increase of energy production in developing countries (AEO 2009), especially in India and China (LeQuere et al., 2009). The challenge to reduce carbon emissions in the developing world is to achieve the targets without placing the legitimate development goals at risk. A distinct, but internationally recognized, action for greenhouse gases (GHG) mitigation in developing countries is the Nationally Appropriate Mitigation Actions (NAMAs). In Brazil a recent government communication to the United Nation Framework Convention on Climate Change (UNFCCC) states as a national mitigation action the "increase in energy supply by hydroelectric power plants, with estimated reduction of 79 to 99 million tons of $CO_{2\text{-eq.}}$ in 2020".

According to the International Energy Agency (AEO, 2009) the projected demand for electrical energy is estimated to increase, by 2030, 85% over 2004 consumption, with higher participation from developing countries, posing a big pressure on the production systems. The hydropower generation is as an attractive renewable energy supply, especially in regions where natural water resources are abundant (WEO 2007), however questions were raised in the past years concerning the production of GHG by these systems, especially in tropical regions, which in terms of climatic change it could be over sighted.

Considering the global carbon cycle, the ecological functioning of hydropower reservoirs became focus of extensive scientific investigation, once emissions of GHG, especially methane, brought concerns about the contribution of these systems to the regional and global carbon balance (Giles, 2006). We identify in the literature detailed studies of the carbon budget focusing on lakes, rivers and other freshwater ecosystems, although only few data related to man-made systems (St Louis et al., 2000, Rosa et al., 2004, Santos et al., 2009). The lack of detailed and complete budget involving carbon inputs to the reservoirs and release to the atmosphere might misjudge the contribution of large hydroelectric reservoirs to the global GHG balance.

A complex and integrated evaluation of several Brazilian tropical hydroelectric reservoirs was proposed in order to understand the carbon cycling and the contribution of these systems to the GHG regional balance. The project named "*Carbon Budget in Hydroelectric Reservoirs of FURNAS Centrais Elétricas S. A.*", was carried out from 2003 through 2008, accomplishing 27 field campaign distributed in 8 reservoirs (see Methods). The literature review by the time when the study was proposed reveal that the information on GHG emission from large tropical hydropower reservoirs were restrict to punctual information, mainly measuring emissions from the water surface to the atmosphere (Galy-Lacaux et al., 1997, Lima et al., 1998, Mativienko et al., 2000, Duchemin et al., 2000, St. Louis, 2000). The study described here brings observations on the carbon input and output; biological dynamic; water flow and physical dynamic in the reservoirs; watershed drivers of changes in carbon flow and budget.

1.1. Context

The human evolution on Earth describes a growing trajectory on using natural resources. After the Industrial Revolution (1750), the Earth atmosphere experienced a gradual and persistent increase of gaseous species derived from processes potentiated by Humans. The emission of these gases, known as greenhouse effect gases (mainly represented by carbon dioxide - CO_2, methane - CH_4 and nitrous oxide - N_2O), brought their concentration in the atmosphere to be in a level non-analogue for the last million years (IPCC, 2007). Changes in atmospheric concentration of an element, molecule or particle, are mainly due to the balance between emissions and sinks, being the unbalance of these vectors the subsidy to the change in concentration.

According to the UN-Millennium Ecosystem Assessment (www.maweb.org), the changes in land use for agriculture altered circa of 25% the continental surface. The land use conversion to agriculture and pasture was higher during the second half of the 20th century than the previous 200 years. According to Cerri et al. (2009) and La Rovere e Romeiro (2003), almost 2/3 of the Brazilian emissions are associated to change in land use and agriculture, and the other 1/3 to transport and energy production. Great deal on that refers to strong participation of renewable source of energy in the country. Hydroelectricity can account for up to 80% of the country needs. The production of hydroelectricity alters substantially the land cover, by damming a lotic system and flooding a big portion of the land and the associated vegetation. On this conversion rely most of the concern related to the GHG emissions, especially methane (CH_4), from reservoirs. On top of this, dams, drastically changing the ecosystem dynamics, fragment 60% of the rivers in the world. Therefore, the clear understanding of the impacts and the social benefits of the energy produced ought to be considering on planning the expansion, and implementation, of hydropower reservoirs.

The global significance of reservoirs as sources of greenhouse gases depends on the total surface area of reservoirs and the flux rates from the major types of reservoirs in different geographical locations (Rosenberg et al. 1997). However the contribution of hydropower reservoirs to the annual anthropogenic GHG emissions might not be negligible. According to Saint Louis et al. (2000) the global area covered by large reservoirs can be at the order of 1.5 million km^2 with significant contribution to the carbon emissions of 7% of the other anthropogenic sources. Thus, taking in consideration the potential of anthropogenic energy production in global climate change (IPCC 2007), the balance between the energy produced and the emission of GHG is critical.

In hydropower reservoirs the emissions of GHG happens, overall, in the lake, in the spillways and in the river downstream the dam. Important information for the GHG balance in these systems refers to the background emissions in the area prior to the lake formation. The high complexity of the ecosystem functioning and carbon dynamic, in the land and river, prior to the dam determine a high level of uncertainty to the post-damming estimations. Therefore, defining the carbon budget in these systems is not straightforward. An ideal situation would be long-term observation studies before and after the dam is constructed. In this way natural fluctuation of the system (i.e, climatic seasons; extreme events, etc) is buffered by the series of observation.

2. Study Site and Methods

In a wide range of age, size, and shape, seven reservoirs are located in central and southeastern Brazil in a biome known as Cerrado (Figure 1) and one (Funil) is located in an highly industrialized region in the Atlantic forest region. Samples were taken during three climatic seasons in each reservoir: at the beginning of the rainy season (November), at the end of the rainy season (March-April), and during the dry season (July-August). The reservoirs Furnas, Mascarenhas de Moraes and Luiz Carlos Barreto de Carvalho (L.C.B. de Carvalho also named Estreito) are in cascade with Furnas upstream and L.C.B. de Carvalho downstream. Corumbá is immediately upstream from Itumbiara, however smaller in size. The reservoirs Manso and Serra da Mesa do not have any damming upstream, on any midsize river within the watershed. Cerrado biome represents 22% of Brazil, with an area of 2.3 million km^2. The average annual precipitation in the region varies from 1200 to 1800 mm, with about 90% of the precipitation concentrated between October and March resulting in two very distinct climatic seasons (dry and wet).

An Environmental Monitoring System (SIMA), an anchored buoy system moored near the dam, was used to provide high frequency environmental data. This system is composed by a set of hardware and software projected for data acquisition, and real time monitoring of hydrological systems (Stech et al., 2006), with data storage systems, sensors (air temperature, wind direction and intensity, pressure, incoming and reflected radiation and thermistor chain), solar panel, battery and transmission antenna. The data are transmitted by satellite in quasi-real time for any user in a range of 2,500 km from the acquisition point. The environmental parameters measured by the SIMA system were consider taking into account the variables that respond consistently to alterations in the

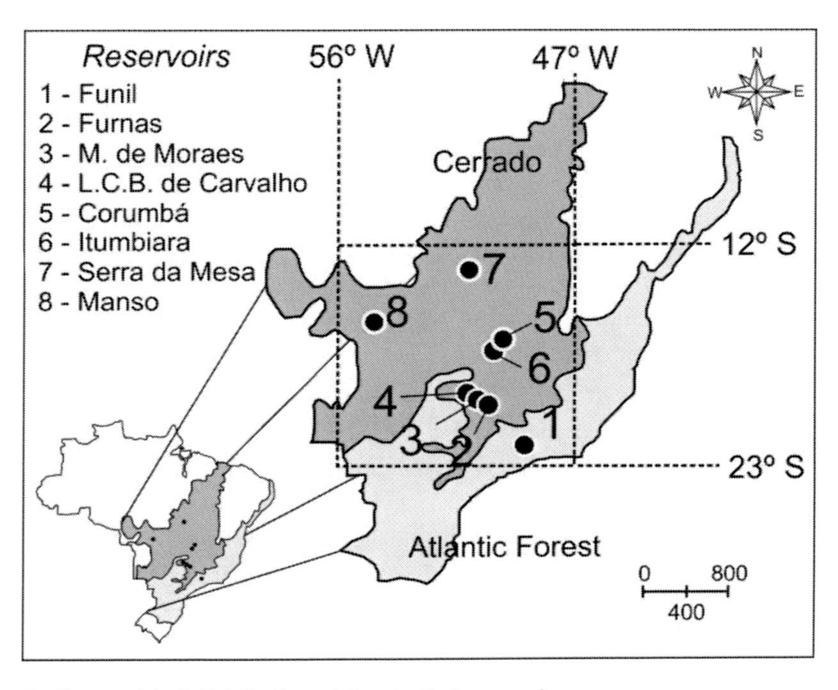

Figure 1. Geographical distribution of the studied reservoirs.

functioning of the aquatic system and its importance for the greenhouse gas emission process in aquatic systems. The carbon load from the catchment to the reservoirs was estimated by the discharge and the concentration of total dissolved and particulate organic carbon in several inflowing streams. Discharges from all major tributaries to the reservoirs were estimated, several times during the field campaigns, through an acoustic Doppler profiler (ADP, River Surveyor – Sontek). The outflow data, as part of the operational routine, was provided by FURNAS CENTRAIS ELETRICAS (www.furnas.com.br). Total Organic Carbon plus Dissolved Inorganic Carbon were determined in samples taken at the same sites where discharge measurements were performed. The water samples were analyzed by combustion and infrared spectrophotometric (TC Analyzer Shimadzu). The mass of input carbon from the tributaries and output downstream the dam ($g\ m^{-2}\ d^{-1}$) in each reservoir, was determined by multiplying the average carbon concentrations ($mg\ L^{-1}$) by the rate of water flow ($m^3\ s^{-1}$).

For the pelagic metabolism processes, each system was sampled in an epilimnion vertical profile in one station near the dam (triplicates samples), at five depths according to light extinction (100%, 75%, 50%, 10% and 1%). For phytoplankton production clear and dark bottles were spiked with 6 μCi of

$NaH^{14}CO_3$. Two tubes were fixed with 4% formalin to abiotic control. After 4 hours of incubation, samples were filtered through 0.45 μm Nucleopore filters, and particulate material was examined by liquid scintillation counting. Phytoplankton production was estimated using radiocarbon technique according to Wetzel and Likens (1991). Phytoplankton and bacterial respiration were measured by dissolved oxygen consumption experiments (24 hours dark incubation). Samples were taken at the same 5-depth profile and filtered in a 68 μm net in order to remove planktonic grazers. After, samples were filtered in 0.7 μm Whatmann filters to estimate bacterial respiration. The unfiltered sample originated the total consumption (bacteria+phytoplankton). The initial control samples were immediately fixed with Winkler reagents. After incubation, samples were fixed with Winkler reagents and dissolved oxygen concentration was determined by spectrophotometric Winkler technique according to Roland et al. (1999). The oxygen consumption data were converted to carbon respired assuming a respiratory quotient of 1. Bacterioplankton production were measured using Smith and Azam (1992) microcentrifuge modification of the [3H]-leucine method (Kirchman and others 1985) in all the 5-depth profile samples. The samples were incubated with 59nM final concentration of [3H]-leucine during 45 min. Incubations were ended by adding 0.3% of 50% TCA. Zero-time controls (blank) were fixed with 50% TCA after adding labeled leucine. Following incubation, the samples were centrifuged at 14,000 rpm for 10 min and the supernatant was discarded. Subsequently, 1.5 mL of 5% TCA were added and the samples were centrifuged again. Afterwards, the supernatant was suck out and a scintillation cocktail was added. The centrifuge tubes were placed in scintillation glass vials, and the radioactivity was determined using a LS 6500 Beckman Coulter. Bacterial production was calculated from disintegration per minute (dpm) to protein according to Simon and Azam (1989).

Adapted from Santos et al. (2004) methodology, bubbles emissions and diffusive gas exchanges at the water–air interface were determined in both upstream and downstream the dams. The ebullitive emissions were determined using 0.38 m^2 funnels placed 30 cm below the surface in several places in the reservoirs in depths varying from 5 to 20 m. Funnels were deployed at each sampling station for 24 hours. Emissions were interpolated to the whole reservoir by weighting for reservoir morphometry. At the water–air interface gas exchanges were evaluated using 1000 ml diffusion chambers sub sampled at a series of times (0, 3, 6, 12 minutes), also adapted from Santos et al. (2004). Samples in the outflow were collected until 1 km downstream the dam. The emissions associated to the turbines operation were not considered in our calculation.

A gas chromatograph (GC) equipped with a thermal conductivity detector (TCD) and flame ionization detector (FID) was use to measure CO_2 and CH_4 in the samples. The system was calibrated in 100% humidity and a 63.6% CH_4 standard, in addition to a commercial 763 ppm CO_2 dry standard. Gases were separated at 26° C in a 3 m long, 1/8 diameter stainless-steel packed column, filled with Hayesep mesh (D 80–100; Vici Valco Instruments, Houston, TX, USA). The carrier gas was hydrogen, at a flow rate of 53 mL min^{-1}.

For the spatial variability of partial pressure of CO_2 (pCO_2), direct measurements of pCO_2 were made approximately every 1,000m along the length of five reservoirs (Furnas, Mascarenhas de Moraes, L.C.B. de Carvalho, Manso and Funil). Triplicates for each sampling point were taken at 0.5 m depth between 8 and 12 pm. The direct measure of pCO_2 was through headspace equilibrium with ambient air (Hesslein et al. 1991; Cole et al. 1994).

The headspace gas was transferred to a plastic syringe, and the concentration of CO_2 in the headspace gas was immediately measured using an infrared gas analyzer (IRGA – environmental gas monitor EGM-4; PP Systems). For the extractions, corrections were made for barometric pressure and for the small amount of CO_2 introduced during the headspace equilibration. CO_2 flux was estimated from pCO_2 according to Cole and Caraco, 2001.

The methodology used by Abe et al. (2005) for sediment sampling was adopted in this study. A sediment corer (UWITEC, Austria), adapted to 63 mm diameter tubes and 600 mm long, was used to collect porewater gas samples. Carbon, nutrients, and porosity were determined in a second core, by slicing the core sample. The sediment cores were processed within 2–3 h after collection, and the gases measured within 4 h after core processing. The gas concentrations in the sediment pore water were expressed in mmol L^{-1}. Sediment subsamples were dried at 70°C until constant weight for percentage water, and further measured by loss on ignition (LOI) at 550°C for 2 h. Gas concentrations were measured as described above. The organic carbon in the sediment was measured using a carbon analyzer (Shimazu, model SSM 5000-A). Fick's first law (Equation 1) was used to calculate the diffuse fluxes of porewater gases to the sediment-water interface, from both the near surface (0.5–1.5 cm) and deeper (1.5–10 cm) sediment layers (Abe et al. 2005; Adams 1994):

$$J = -\varphi Ds\left(\frac{dc}{dz}\right)$$

Where, J = the diffusive flux (g m^{-2} d^{-1}); φ = porosity; Ds = sediment diffusion coefficient for each individual gas (m^2 s^{-1}); and dc/dz = the concentration change of each gas with depth (g m^{-3} d^{-1}).

The water-atmosphere interface and the energy generated (MWh) by each reservoir, provided by FURNAS, are routinely calculated, at operational basis, from the water level of the reservoir. The $CO_{2\text{-}EQ.}$ (carbon dioxide equivalent) was calculated by adding the data of CO_2 plus the methane (considering the global warming potential of 21) based on Lelieveld et al. (1998).

3. Results and Discussion

The carbon dynamics in hydroelectric reservoirs involves three major compartments – sediment, water and atmosphere – and processes of production, storage and release of carbon (Figure 2). This section brings the main results of the project "Carbon Budgets in Hydroelectric Reservoirs of FURNAS Centrais Elétricas S.A.".

The results are organized in four topics according to the pathways presented in Figure 2, as follow: (1) input and output by the rivers; (2) sediment carbon concentration and fluxes (3) flux in water – air interface and (4) processes in water column. Then, the principal factors controlling the fluxes of greenhouse gases from reservoirs are discussed, also based on results from that project.

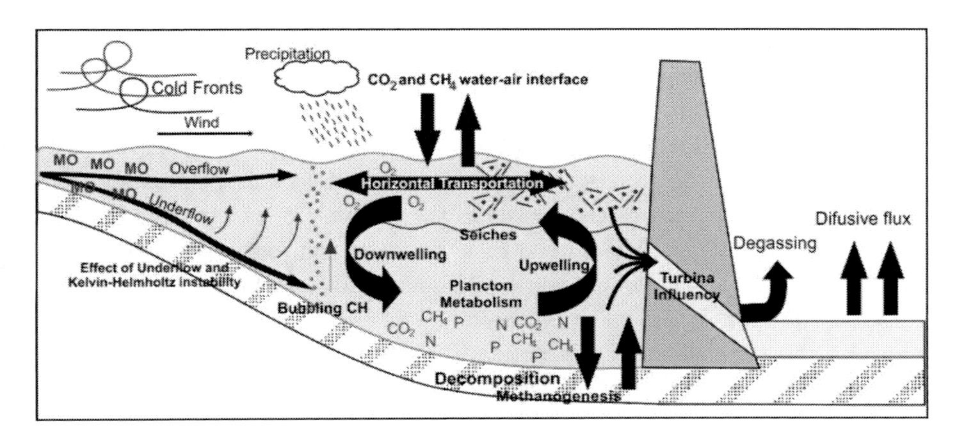

Figure 2. Major carbon pathways in hydropower reservoirs.

3.1. Input and Output by the rivers

In artificial reservoirs, climatological (e.g. precipitation, wind, solar radiation and temperature) and other physical forcings derived from the operational routine of the system (e.g. residence time, water discharge, level fluctuations, advection currents) lead to complex hydrodynamic patterns. Such patterns reflect in the biological dynamic of the reservoir, as, for instance, the spatial and temporal distribution of planktonic organisms (Tundisi, 1990). Aside the physical drivers, the input of organic matter produced in the drainage area (allochthonous matter) represents an important regulator of the reservoir biology.

The close connection between aquatic and terrestrial systems was highlighted in the literature (Cole and Caraco, 2001), indicating the importance of terrestrial material to the aquatic metabolism. Nevertheless, high terrestrial material loads can indicate an inadequate management of soils within the drainage basins. The Cerrado region, in particular, is characterized by sand and low-fertile soils, which requires a conservative, non-tilling, agricultural production. Additionally, in Cerrado, degraded pasture represents more than 60% of areas for cattle production (Mercedes Bustamante et al., "Estimating Recent Greenhouse Gas Emissions from Cattle Raising in Brazil", in review), therefore under high risk of erosion and loss of soil organic matter.

In the scope of the project, the contribution of allochthonous carbon to the organic matter pool of the reservoirs was estimated through carbon stable isotopes analysis. This study was carried out in the reservoirs: Serra da Mesa, Manso, Corumbá and Itumbiara. The contribution of allochthonous material and its distribution within the reservoirs were distinct among the systems. The intense changes in land cover in the drainage basins of Corumbá and Itumbiara reflected in a high load of allochthonous material to the reservoirs. In Corumbá, the influence of non-treated domestic sewage was also detected. Contrastingly, the importance of patches of preserved area in the drainage area was highlighted in Serra da Mesa and Manso reservoirs, once the terrestrial organic matter load in these systems was low in comparison to Corumbá and Itumbiara. Despite the fact that the contribution of allochthonous material was distinct among the reservoirs, our results suggest that this is an important rout for carbon load to all studied systems, as illustrated in Figure 3 for Manso reservoir. The output of carbon to rivers downstream the dams were also high. In the dry season sampling the carbon output downstream exceeded the input from rivers in Serra da Mesa, Itumbiara, Manso, Corumbá and Furnas, suggesting that internal sources of carbon (i.e., gross primary production, flooded biomass decomposition) play a significant role in the reservoirs' carbon balance (Figure 4). The input of carbon, in rainy months, exceeded the downstream output for most reservoirs (Figure 4), suggesting a strong contribution from the watershed and important sedimentation in the reservoir.

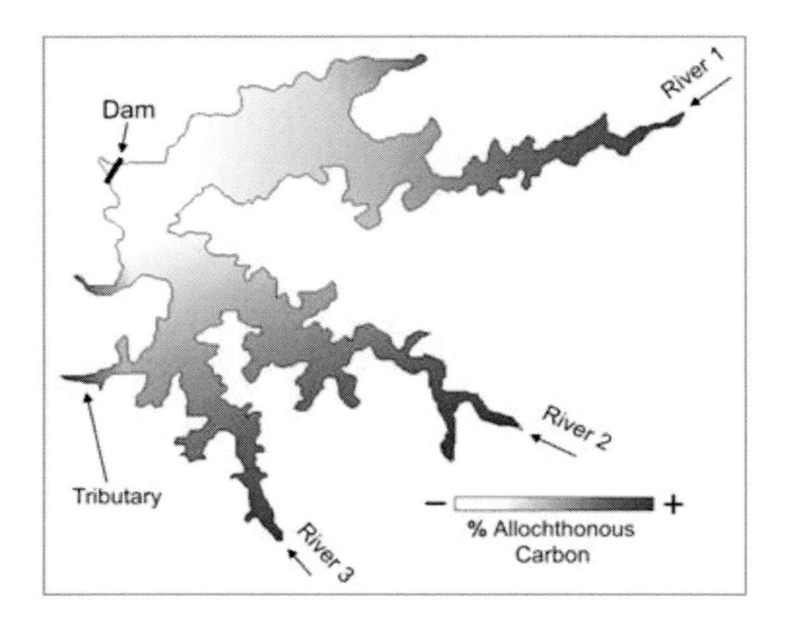

Figure 3. Schematic representation of the allochthonous carbon contribution to Manso reservoir.

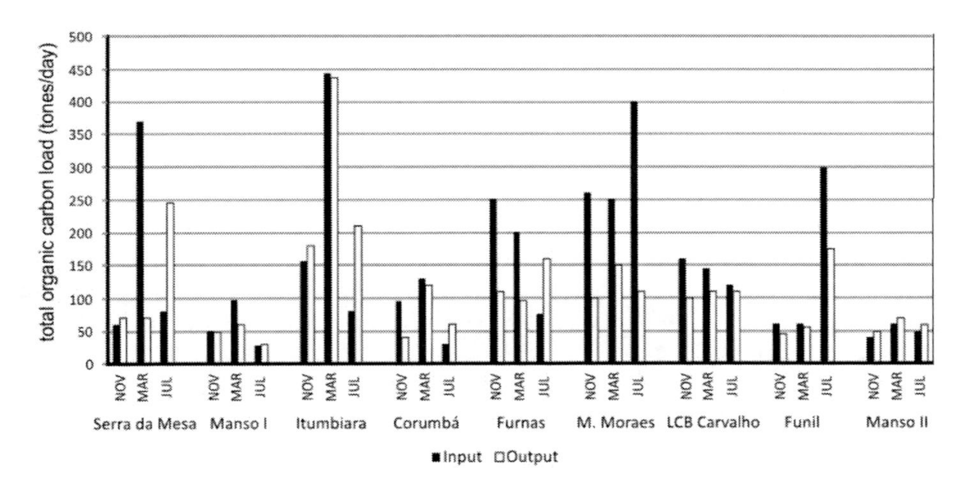

Figure 4. Total organic carbon load (tons/day) considering the input upstream and output downstream of the reservoirs.

In the cascade reservoirs (Furnas, Mascarenhas de Moraes and Estreito), the output of organic carbon downstream was lower than the input of carbon from rivers upstream the reservoirs. This observation suggests that those three reservoirs may represent a carbon sink. However, the allochthonous carbon could create heterotrophic condition, resulting in net emission of CO_2 and CH_4 to the atmosphere.

3.2. Sediment Carbon Concentration and Fluxes

Sedimentation is a key process in reservoir carbon dynamics (Dean and Gorham, 1998). The deposition of organic carbon in reservoir sediments is known as "fresh sedimentation". Once in the sediment, the deposited organic carbon is either mineralized by heterotrophic organisms or remains untransformed into the sediment. The mineralization process produces CO_2 and CH_4, which diffuse to the water column, while the carbon permanently accumulated in the sediment builds the carbon stock in this compartment. Sikar et al (2009) concluded that the permanent carbon sedimentation in 7 out of the eight studied reservoirs – Funil was not considered in this study – removed significantly more carbon the soils in an equivalent area, suggesting reservoirs in this region as large carbon sinks. In Manso reservoir, the youngest among the studied systems, two sampling campaign were planned. During the first and second sampling periods (2003-2004 and 2006-2007), approximately 8% of the fresh sediment was trapped in the sediment as permanent carbon, regardless of the season. In the other reservoirs the fresh/permanent sedimentation rate varied considerably over the months. The highest percentage of permanent carbon sedimentation was observed in Mascarenhas de Moraes and Luiz Carlos Barreto de Carvalho: about 40% of the carbon reaching the bottom sediment of the reservoirs was retained as permanent carbon.

We identified no relationship between fresh or permanent sedimentation and greenhouse gases emissions in the reservoirs. The sedimentation rates observed in our study (32 to 68 gC m^{-2} y^{-1}) are comparable to values obtained for other tropical lakes (10.8 to 250 gC m^{-2} y^{-1}, Smith et al. 2003, Moreira-Turcq et al. 2004, Devol et al. 1984, Smith-Morrill 1987). For temperate lakes, Kortelainen et al. (2004) registered a mean sedimentation rate of 1.76 gC m^{-2} y^{-1}, rather low when compared to tropical systems. A general trend of decreasing CO_2 concentrations in the sediment with increasing the time after reservoir damming was observed. Manso, the youngest reservoir, had the highest carbon concentration in the sediment in 2003-2004, among all reservoirs (Figure 5a).

Manso was sampled again after three years (2006-2007), when significant increase in CH_4 and reduction in CO_2 concentrations in the sediment were observed. These observations allowed a temporal evaluation of the concentrations and diffusive fluxes of CO_2 and CH_4 in the sediment.

According to Abe et al. (2005) the increase in organic matter content and nutrients (nitrogen and phosphorus) in sediments during the early period after damming and a consequent decrease in dissolved oxygen concentration and redox potential in the water above the sediment-water interface would be the major drivers for the gases concentration behavior observed. The mean oxygen concentration and redox potential for both sampling periods (2003-2004 and 2006-2007) were respectively 0,96 and 0,36 mgL^{-1} and 174,5 to -13,0 mV. The changes to a reduced environment, from the first to the second sampling period, favor the CH_4 in relation to CO_2 production in Manso sediments. The gases fluxes from the sediment pore water to the water column depend on a suite of environment physical conditions in the hipolimiun and in the sediment. Thermal stratification of the water column, anoxic conditions, trophic status of the water body and concentration of organic matter in the sediment are important drivers

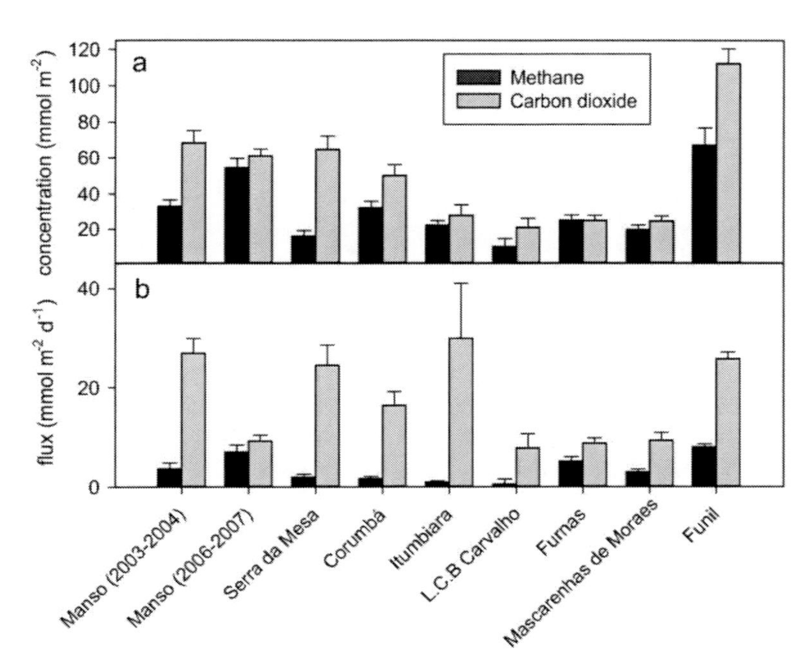

Figure 5. Carbon in the sediment. a)CH_4 and CO_2 concentrations b) CH_4 and CO_2 flux in the sediment-water interface.

contributing to CH_4 fluxes in sediment-water interface. Higher CH_4 fluxes observed in Furnas, Mascarenhas de Moraes and Funil reservoirs are associated to favorable conditions of the physical drivers described above. For the other reservoirs an inverse relationship emission/age was observed for both CH_4 and CO_2 (Figure 5.b).

The trophic state of Funil also contributed to higher concentrations of both CH_4 and CO_2 in the sediment. Liikanen et al. (2002) also observed higher biogenic gases concentration in eutrophic water bodies, which reflected the higher biomass and organic compounds that contribute to the stock of carbon in the sediment.

In general, the diffusive fluxes of CH_4 and CO_2 observed in reservoirs are higher than values observed in natural lakes of similar trophic status (data not shown). The difference in flow between reservoirs and lakes may be related to the intense input of allochthonous material to reservoirs, resulting in greater accumulation of organic matter in the sediment of these systems.

4. Processes in Water Column

Besides the fluxes between reservoir compartments, the processes involving production, consumption and stocks of greenhouse gases within the water column were approached in the project. Dissolved CO_2 can be consumed through photosynthesis; produced by aerobic respiration or stored as dissolved inorganic carbon (Kalff, 2001).

On the other hand, CH_4 can be produced in anaerobic regions by methanogenic bacteria or consumed through oxidation by methanotrophic bacteria at the interface between oxic and anoxic regions. The data on production and consumption of CO_2 (primary production, phytoplankton respiration, bacterial production, bacterial respiration and bulk plankton respiration) are presented in Table 1.

Primary production in the water is controlled by a variety of limiting factors. In reservoirs the main factors limiting primary production are nutrients concentrations (phosphorus and nitrogen) and light. Our study showed that, unlike expected, eutrophic environments, where nitrogen and phosphorus are abundant, are not the most productive ones. For example, despite being an eutrophic system, primary production in Funil reservoir is highly limited by light.

The high phytoplankton biomass in this reservoir obstructs light penetration into the deeper layers. Serra da Mesa, on the other hand, had the highest primary production rates among all sampled reservoirs. Serra da Mesa is characterized by

Table 1. Plankton metabolism in Manso, Serra da Mesa, Corumbá, Itumbiara, Furnas, Mascarenhas de Moraes, Estreito, Funil e Manso

Reservoirs	Primarie Production		Phytoplakton Respiration		Bacterial Production		Bacterial Respiration		Plancto Respiration	
	mg C m^2 dia^{-1}									
	Average	DP	Average	DP	Average	DP	Average	DP	Average	DP
Manso	113	47	346	76	240	86	1036	208	1382	284
Serra	214	112	278	58	206	155	821	268	1099	326
Corumbá	25	16	381	187	50	61	552	501	932	661
Itumbiara	189	44	475	307	56	1	651	255	1127	99
Furnas	54	26	319	58	57	5	397	216	716	200
M. Moraes	18	4	208	56	61	18	405	131	613	187
Estreito	39	5	280	166	55	10	555	284	834	449
Funil	46	39	153	68	291	307	457	157	610	198
Manso	34	11	143	117	111	11	448	490	592	605

low turbidity and nutrient levels above those considered limiting to phytoplankton growth. Low rates of primary production were also observed in Corumbá, which can be explained by the high inorganic turbidity and consequent low light penetration depth.

Net ecosystem production (NEP - primary production minus ecosystem respiration; negative values denote heterotrophy while positive values denote autrotrophy) was inversely related to the age of the reservoirs. This pattern suggests that fresh flooded biomass represents an extra source of energy for respiration in young systems.

Manso data, sampled in two periods (2003-2004 and 2006-2007) demonstrate the influence of age in the reservoir NEP. Funil was the only reservoir with positive NEP, meaning that production exceeded respiration in this system.

5. Flux in Water – Air Interface

5.1. Diffusive Flux

The gas exchanges in the water-atmosphere interface are an important integrative property in reservoirs, reflecting many processes in the aquatic systems (rivers, reservoir) and in the watershed. The geometry of the surrounding and bottom of tropical reservoirs are associated to complex mosaics of land-cover and land-use, which defines distinct patterns of lateral contributions and distribution of flooded biomass. Patterns that reflect in the spatial distribution of GHG fluxes in the studied reservoirs.

The determination of greenhouse gases fluxes in the water-air interface of freshwater systems is not a straightforward task. Many uncertainties are associated to equipments and sampling strategies, but also to meteorological aspects.

For example, flux measurements using static floating chambers provide relatively short timescale data (1-3 hours). However, recent laboratory experiments indicate that static floating chambers can provide reliable results when they are used under moderate wind conditions (Kremer et al., 2003, McGilid and Wanninkhof, 1999).

The CO_2 emission in tropical hydroelectric reservoirs (mainly in Amazon and Cerrado regions) is higher than the CO_2 emission in temperate hydroelectric reservoirs. This pattern is probably associated to topographical conditions and the higher amount of flooded biomass in the Tropical regions (Barros et al, "*Global magnitude and regulation and carbon emission from hydroelectric reservoirs*", in preparation).

5.2. Bubble Flux

The emission of CO_2 bubbles was, in general, small for all studied reservoirs. The flow of methane bubbles tends to be positively related to the amount of flooded organic matter during impoundment and, consequently, negatively related to the age of the reservoirs. For example, in Manso reservoir we observed bubble fluxes about three times higher than diffusive fluxes (on average) for the first period of sampling, reflecting the recent filling of the reservoir. In the second sampling period (after 76 months of impoundment), with reduction in the bubble fluxes, diffusive and bubbling CO_2 transfers to the atmosphere were similar, in unit per area. A similar pattern was observed in Serra da Mesa, where after 72 months from impoundment CH_4 bubble fluxes were about five times higher than diffusive fluxes. These values substantially reduced over subsequent field works (Marco Aurelio Santos, COPPE, personal communication).

Older reservoirs, such as Furnas and Mascarenhas de Moraes, showed small fluxes of bubbles from the sediment. In Mascarenhas de Moraes the bubble fluxes were close to zero. On the other hand, Corumbá and Funil reservoirs showed high CH_4 bubble emissions. The results for Corumbá and Funil might reflect the high input of organic material from the drainage basins and the important load of domestic sewage to these systems (unpublished data).

5.3. Factors Controlling the Flux

5.3.1. Age

The negative relationship between the GHG flux and the age of the reservoirs was a relevant outcome of this project (Jean P. Ometto et al., "*Patterns of carbon emission as a function of energy generation in hydroelectric reservoirs*", in review). The same relationship has been shown for other reservoirs (St. Louis et al. 2000). Our data showed that the relationship with reservoir age was especially strong for CH_4 emission as bubbles in the water-air interface . This result indicates the importance of the flooded organic material to GHG emissions, and suggests an improve in the CH_4 emission/energy production ratio with time.

In general we also found a trend of decrease in the GHG concentrations (CO_2, CH_4 and N_2O) in the sediment over the years. The same pattern was observed for the diffusive flux of CO_2 in the sediment-water interface.

5.3.2. Climate Dynamics

5.3.2.1. Effects of Cold Fronts in the Stratification of Manso Reservoir

Cold fronts that reach the Brazil are usually formed in Southern South America and progress to lower latitudes, reaching the Central part of Brazil and the Cerrado region. This process is usually associated with local temperature, relative humidity and air pressure decrease, which are accompanied by the intensification and change in wind direction from the typical NE/E for SW/S. From early May until July of 2004, five cold fronts reached Manso reservoir. The thermal profiles during the first three cold fronts show the effect on the thermal structure of the reservoir (Figure 6). The thermal profiles also show a transition from a stratified reservoir in rainy season to an almost mixed water column in dry season. During each front, peaks of wind mixing index (third power of wind intensity) were observed followed by air temperature decrease (Figure 6).

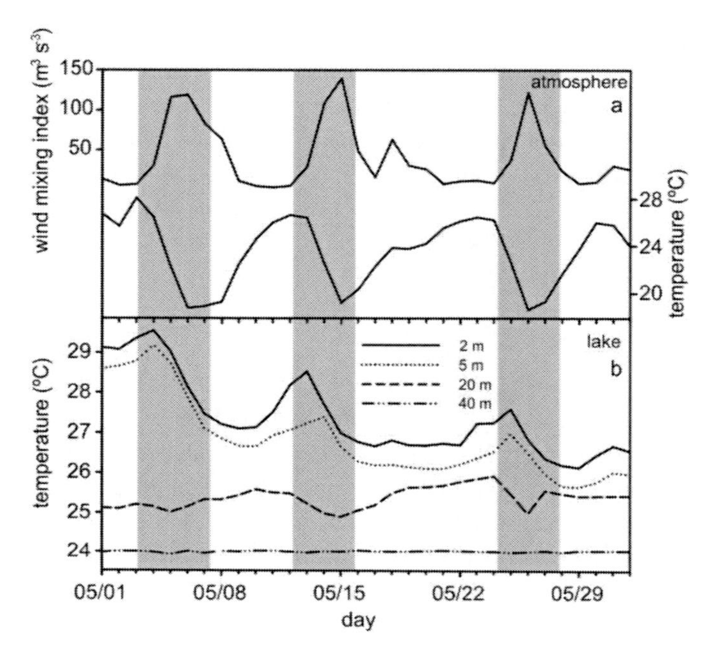

Figure 6. Diary observation of Manso reservoir obtained by SIMA from 01/05 to 01/06, 2004. Gray parts highlight periods of temperature decrease due to cold fronts. The upper panel (a) shows the variation of atmospheric temperature (°C) and wind mixing index (m3 s-1). The bottom panel (b) shows the effect of the cold fronts in the water temperature at 2, 5, 20 and 40 meters deep.

Atmospheric instability is common during the frontal system, since the variation in temperature reaches 7-8°C during the front (against 1-2°C in normal conditions). Decrease of water temperature can occur at up to 20m deep, but the higher decrease occurs at the top 5m layer (Figure 6.a). The increasing difference in temperature between air and water can strongly affect the heat flows on the surface.After the first cold front, while the weather returns to normal conditions, surface water begins to heat up and re-stratify but to a lower plateau. A similar behavior was observed after the next subsequent fronts. The result of these fronts was a decrease in the average temperature of the upper layer of about 3°C and a substantial reduction of vertical temperature gradient and stratification. The difference of 5°C between temperatures closer to the surface (2 m) and in 20 m depth at the beginning of the period was reduced to about 1°C after a cold front. No sign of temperature change was observed in the depth of 40 m, indicating the meromitic nature of the reservoir during this period.

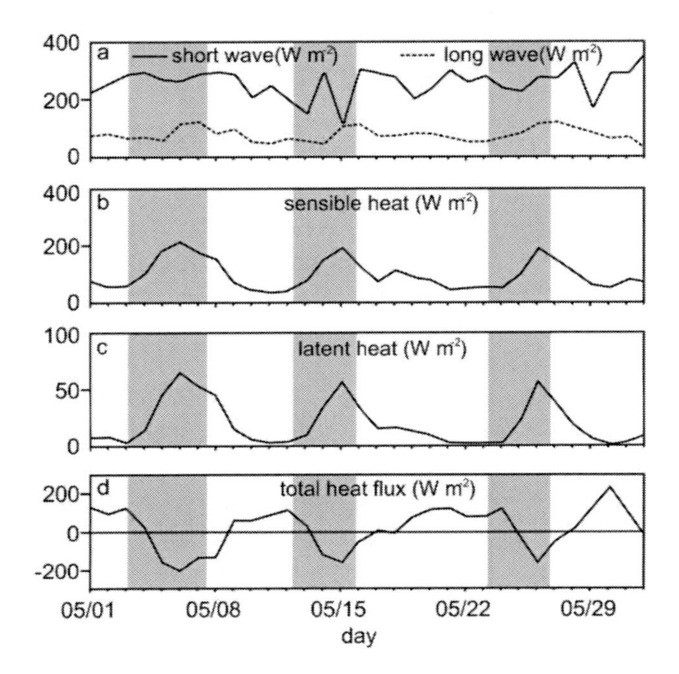

Figure 7. Diary mean of heat flux in the surface water in Manso reservoir, calculated from SIMA data from 01/05 to 01/06, 2004. a) solar radiation of observed short wave and calculated long wave; b) flux of latent heat; c) flux of sensible heat and; d) total surface heat flux. The line in the bottom panel separates periods that reservoir is gaining (positive) and losing (negative) heat.

Figure 7. brings a better understanding of how cold fronts modulate the reservoirs temperature. The development over time of daily heat flow budget on the surface gives a good overview of its relative importance to total net flow. The flow on the surface amplifies changes of all terms. During the cold fronts the main heat loss process on the surface is the evaporation of latent heat, which can reach up to 200 Wm^{-2}. Most of this increase is due to the intensification of wind speed, but part of it is also due to moisture loss. The sensible heat, which is almost negligible in normal conditions, increases up to about 60 Wm^{-2} during the front. These changes are associated with increasing wind and with the difference of temperature in air-water interface. The long wave radiation flow is the heat balance term that is less affected by the front, but its contribution to the total heat loss is relatively important to the average (approximately 80 Wm^{-2}). It is important to note that only few events of cold fronts were enough to change the reservoir from stratified and warm during the spring and summer to almost mixed and cooler in middle autumn.

5.3.2.2. Effects of Cold Fronts in Greenhouse Gases (GHG) Emissions in Corumbá Reservoir

To understand the effect of cold fronts on GHG emission in Corumbá reservoir we used field measures and temporal series of meteorological and limniologic variables in the period of March 12-19, 2005. During this period a cold front affected the reservoir. Besides the direct impact of this front, the entire studied region was under the influence of the South Atlantic Convergence Zone (SACZ). The SACZ is characterized by a persistent and large cloud cover, usually extending from the Amazon region up to southwest of South America. Figure 8 shows the position of the SACZ from 16th to 20th March, 2005. The SACZ which appears at the clearer part of the picture was obtained from meteorological GOES 12 satellite. Similarly to Manso, the wind pulses (Figure 9a) and the decline of air and water temperature (Figure 9c and d) were accompanied by the arrival of the cold front and the development of SACZ on the Corumbá reservoir. The SIMA data showed an increase in the frequency of wind speeds higher than 3 ms^{-1} was more significant than the increased average wind speed after 13 March. The chlorophyll *a*, which was almost non-existent until the front event, increased after 15th March (Figure 9f). Several chlorophyll *a* peaks were observed in this period. A similar pattern was observed to turbidity (Figure 9e). The increasing chlorophyll *a* concentration (possibly produced by the increased availability of nutrients due to the resuspension of sediments in shallow areas in the reservoir) is a clear evidence of water mixing and increase in turbidity levels after the rain.

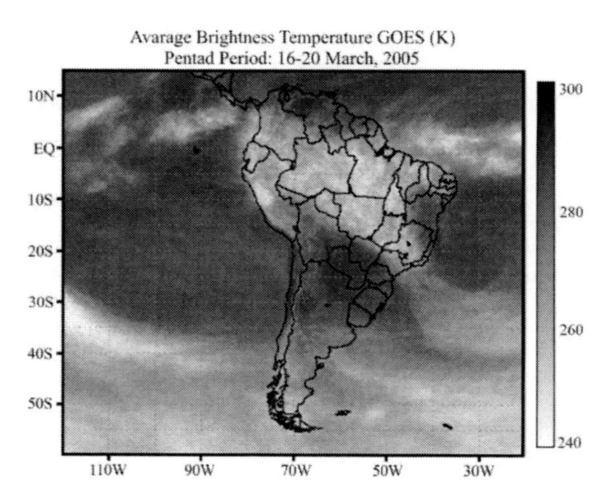

Figure 8. South Atlantic Convergence Zone (SACZ) obtained from GOES 12 satellite average brightness temperature. Period from 16 to 20 March, 2005 (Source: CPTEC/INPE)

.

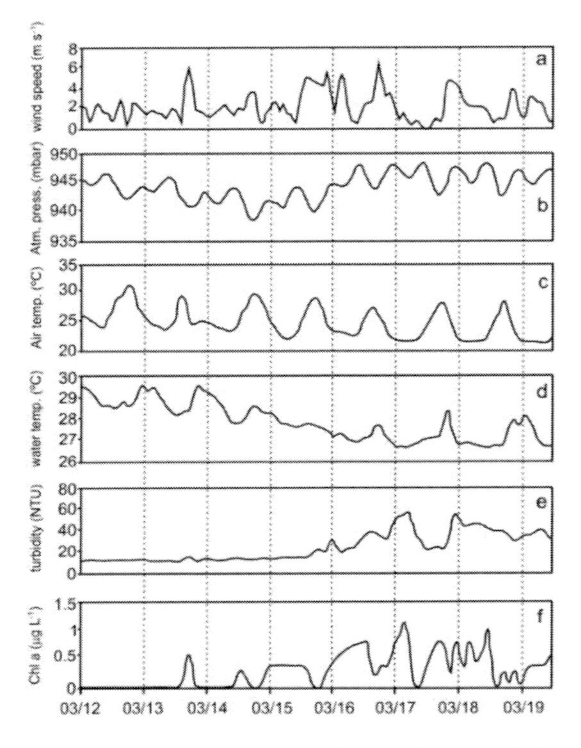

Figure 9. Limnologic data collected by SIMA in the pelagic zone of Corumbá reservoir.

Table 2. Effect of cold fronts in Corumbá reservoir

		After the cold front
Wind Speed	increase of	15%
Turbidity		208%
Chlorophyll a		400%
Carbon Dioxide diffusive flux		244%
Air Temperature	decrease of	8%
Water Temperature		6%
Methane diffusive flux		28%
Methase bubling flux		88%

Table 2. summarizes the effect of cold fronts in Corumbá reservoir for the periods before and after the passage of the front. Bubbling methane that was relatively high before the front passage decreased in intensity and frequency (Lorenzetti, INPE, unpublished data). Figure 9b shows a consistent decline in atmospheric pressure of approximately 6 mbar before the front. Although we tend to assume that changes in emission of CH_4 bubbles could be linked to changes in atmospheric pressure produced by cold front passage, several other conditions are also modified by the front cold. Increases in turbulence and decrease in stratification and mixing of oxygenized water can affect the production of CH_4 and bubble emission.

A persistent reduction of methane diffusive flow was also observed with the arrival of the cold front. The CO_2 diffusive flux increases considerably after the passage of the front. The increase in CO_2 accompanied by a reduction in CH_4 can be caused by the oxidation of CH_4 into CO_2 by metanotrophic bacteria under aerobic conditions.

5.3.2.3. River Power

River temperature varies during the year and in some seasons it is lower (rainy season, December to March) or similar (dry season, May to August) to the temperature in the adjacent reservoir. Because of temperature and other physical characteristics (e.g. total dissolved solids and suspended solids), water from the river and from the reservoir usually differ in density (Ford 1990). Because of this density differences, water currents from rivers can flow into the reservoir in the surface layer (overflow), in the deepest (underflow) or in intermediate layers (interflow) (Martin and. McCutheon, 1999). After the inflow, the river can plunge and follow the original river bed as an underflow. Once an inflow has plunged and formed an underflow or interflow the water mass is isolated from the surface waters. However, Chen et al., (2006) have recently observed that, in fact, the water mass can entrain constituents (dissolved and particulate elements) into surface waters.

Figure 10. shows the river behavior in Manso reservoir during the rainy season. A water plunge occurred 35 km upstream the dam (Figure 10a and d). However the water characteristics at the plunge point were similar to those registered at the hypolimnion near the dam (Figure 10b and c). This observation suggests that the river flowed through the reservoir as an underflow. In other words, the river inflow behaves as a dense fluid that enters and flows under a lighter fluid (Assireu et al., 2011). Due to shear velocity, waves arise in the currents interface and grow towards downstream. The formation of waves is an

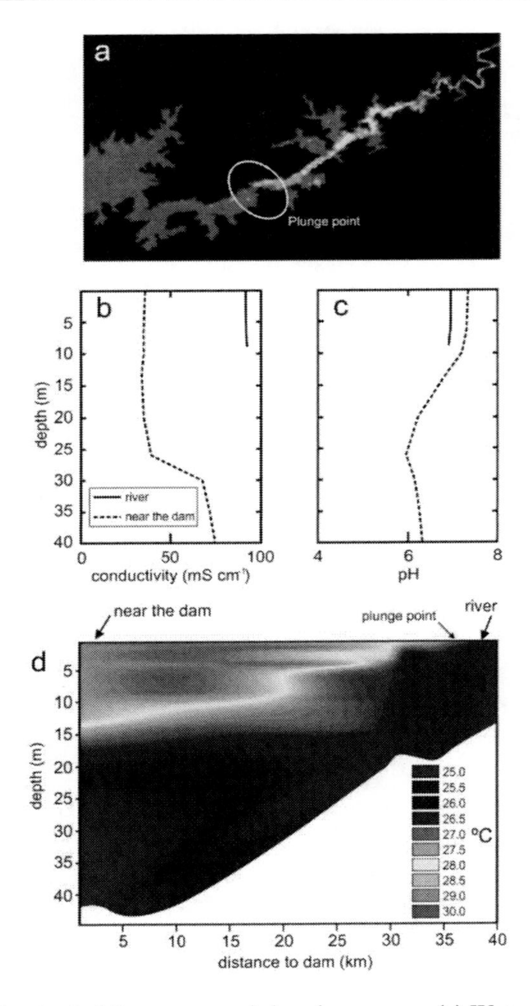

Figure 10. River behavior in Manso reservoir in rainy season. (a) Water surface temperature estimated from thermal band (band 6) of Landsat-5-TM. White circle indicate plunge point. (b) Conductivity and pH vertical profiles at Manso reservoir measured in river and dam station during rainy season. (c) Isopleths of the water temperature in Manso Reservoir in rainy season (Modified from Assireu et al., 2011).

indicative of the Kelvin-Helmholtz instability. This instability causes water movements that dislocate the thermocline up and transport nutrients and gas from deep to upper water layers. Water transport by physical process can break the stability and displace the water layer near sediment. This process causes turbulent transport of gases out of the surface layers of sediment to the water. Changes in the physical environment are critical for the dynamics of matter and energy flows

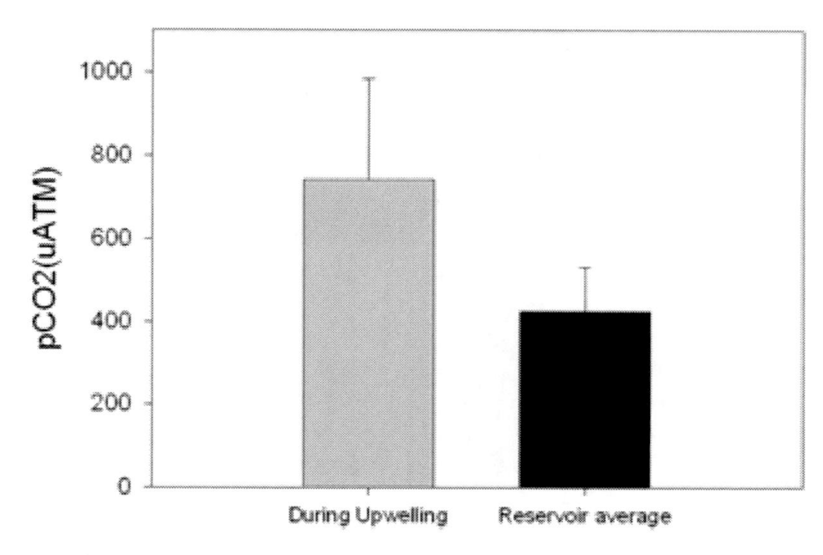

Figure 11. Surface pCO_2, at the sampling station close to the dam (Manso reservoir), during an upwelling event (July 2007) and the reservoir pCO_2 average.

in ecosystems. Consequently, changes in the physical environment affect the metabolism of the system. Thus the integrated study of the physical components that involve the river behavior brings scientific advances for the study of reservoirs.

During the upwelling event study at Manso reservoir, the surface CO_2 concentration at sampling station close to the damwas higher than the spatially averaged CO_2 concentration in the entire Reservoir (Figure 11).

During the upwelling pCO_2 nearly doubled the mean pCO_2 values. We attribute this high surface CO_2 concentration to vertical transport engendered by upwelling, which would pump to the surface the originally stored CO_2 and the CO_2 resulting from methane oxidation.

5.3.2.4 Wind Power

Meteorological events like storms and thermal winds play a crucial role in the gas fluxes in reservoir (Abril et al. 2005), promoting a multiplicative effect on the instantaneous atmospheric fluxes (Roland et. al 2010). In Manso reservoir, mixing events and increase of gas concentration at the surface layer followed a series of strong winds.. Peaks of rain and especially sudden increases in wind speed promoted mixing of the water column.

Figure 12. shows the spatial distribution of pCO_2 before (Figure 12a) and after (Figure 12b) a mixing event in Manso reservoir, which caused deepwater

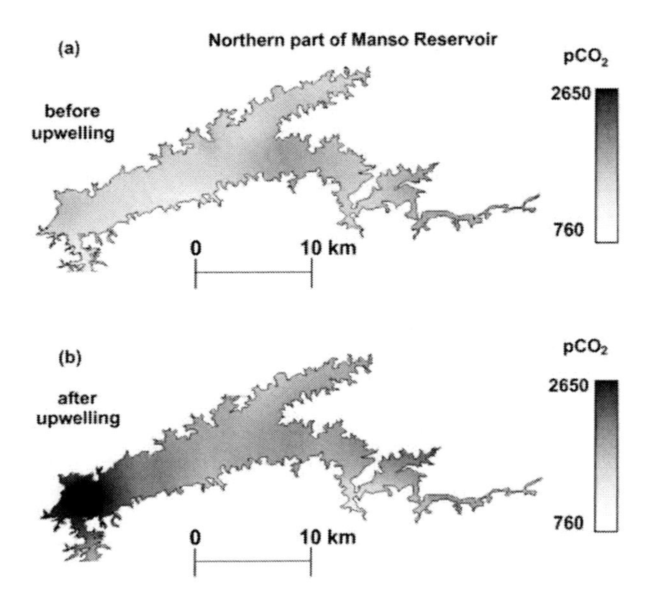

Figure 12. Manso reservoir and pCO_2 data (μatm) during the dry season. Spatial distribution of pCO_2 before (a) and after (b) a mixing event. The gray gradient on both panels represents themagnitude of p CO_2 data using the Ordinary Kriging statistical procedures (Modified from Roland et. al 2010).

upwelling after two days of strong and persistent wind. The values in the region near the dam were seven times larger than the values before the event. According to Assireu et al. (2005) similar phenomena of strong and persistent wind occur in about 10% of the days on July.

Considering this frequency and knowing that the percentage of the area of upwelling influence is 20% of the total reservoir area, CO_2 emission on July is 12% greater than the values calculated for the same period without considering the phenomenon of upwelling. Thus, advances in this approach are critical to an integrated understanding of GHG emissions in reservoirs.

5.4. Spatial Variability

Estimating gas emissions from larger reservoirs is complicated since they are morphometrically complex and spatially heterogeneous. In addition, the spatial variability of GHG emission in large impoundments is dependent on flooded biomass type, watershed input and dam operation. Most currently available data about gas emission have not considered the spatial variability and continuous time series measures at one site can super-or-underestimate emissions. Furthermore, these reservoirs can receive more terrestrial carbon from tributaries along their lengths, what can influence in the total emissions. Roland et al. (2010) show that, considering only data near the dam, as in most reservoir studies, up to 30% of gas flux is underestimated, since the site closest to the dam tended to have the lowest p CO_2 saturation compared to other sites.

Like others, reservoirs located in Brazilian tropical Cerrado are predominately CO_2 supersaturated. However, CO_2 flux to the atmosphere is considerable smaller than those reported for many others reservoirs and, in some cases, sub-saturation of CO_2 was observed in the main body of some system. In general, higher values of emission were found in regions of tributary inputs influence, with decreasing emission toward the dam and increase in downstream, after turbine outflow. The land use around the reservoir (city, farms and livestock) also influenced gas emission. In cascade reservoirs, values of CO_2 emission along main body of reservoir were influenced by turbine water from upstream reservoir rich in CO_2 and CH_4.

Figure 13. shows pCO_2 spatial variability within Furnas reservoir in a spatial scale expressed by a color gradient obtained from an interpolation of measured data using the Ordinary Kriging statistical procedures. The results of spatial variations for all reservoirs studied by Roland et al. (2010) showed some patterns that can be pointed out:

- Areas close to the riverine inflow had higher levels of pCO_2 than the main body of the reservoir;
- pCO_2 decreased towards the dam and increased downriver of the dam for all the systems;
- The pCO_2 spatial variability changed as a result of physical event associated to water vertical mixing movement from the bottom to the surface;
- The highly elevated pCO_2 at some sites in UHE Furnas and UHE Funil co-occurred with the locations of significant sewage outfalls;
- Cascade reservoirs received CO_2 rich hypolimnetic water from upstream reservoirs. UHE Mascarenhas de Moraes reservoir is located just after the dam of UHE Furnas. The higher pCO_2 near the dam (in the reservoir) of UHE Furnas promoted high downriver degassing in UHE Mascarenhas de Moraes. Similarly, UHE LCB Carvalho is just downriver of the Mascarenhas de Moraes dam and is similarly influenced by the hypolimnetic water from this reservoir;

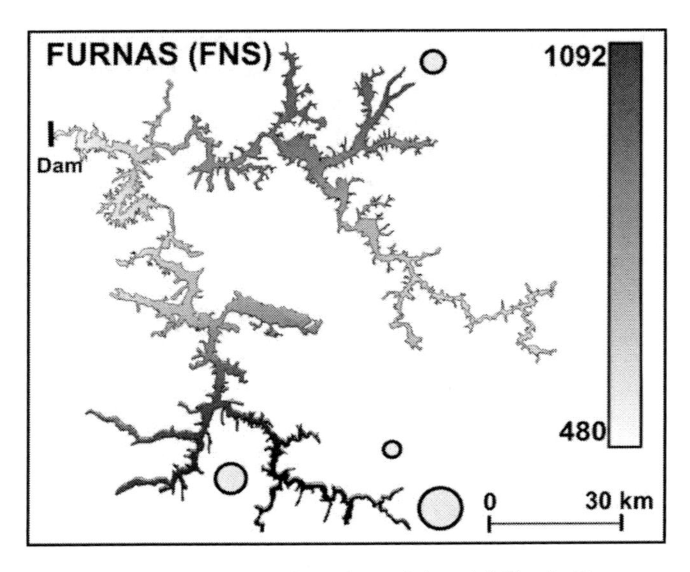

Figure 13. Schematic drawing of pCO2 (μatm) spatial variability in Furnas reservoir. The gray gradient represents the scattering from undersaturation (lightest gray) to supersaturation (darkest gray). Gray circles show the location of the four largest cities around the reservoir; circle sizes represent relative population density (Modified from Roland et al., 2010).

- Variation in reservoir depth causes spatial variation in pCO_2. In UHM Manso, pCO_2 was clearly higher in the shallows and lower in the deep regions. It is likely that variation in depth contributes to spatial variability in the other systems as well.

In summary, a combination of in-system and external forces probably causes the spatial variability in CO_2 flux from hydroelectric reservoirs. Hydrodynamic events, linked to the reservoir depth and water residence, time appears to be key factors modulating the spatial patters of pCO_2. Additionally, external forcing, like watershed dimension and input of carbon, play combined actions explaining the variability.

5.4.1. $CO_{2\text{-}EQ.}$ Fluxes at Water-Atmosphere Interface

The $CO_{2\text{-}EQ.}$ describes, for a given mixture and amount of greenhouse gas, the amount of CO_2 that would have the same global warming potential (GWP), when measured over a specified timescale (generally, 100 years). In our study the GHG mixture was represented by CO_2 and CH_4 obtained from diffusive water-atmosphere interface fluxes plus bubble fluxes of CO_2 and CH_4, extrapolated to

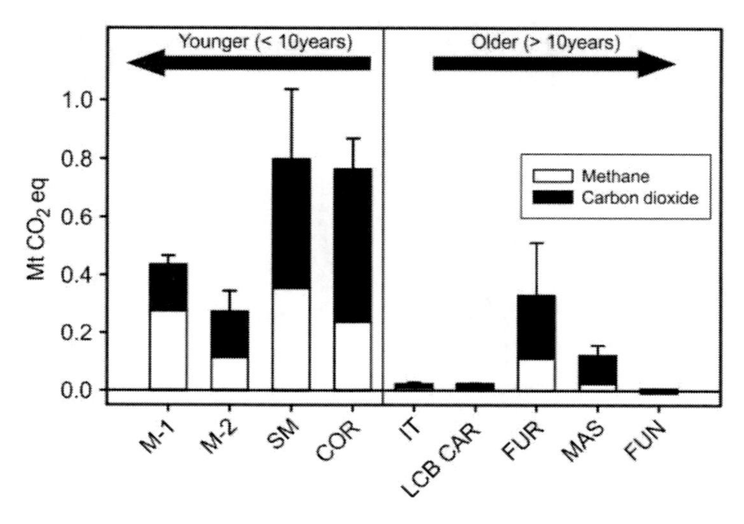

Figure 14. GHG emissions in the studied reservoirs. Manso is represented by two sampling period spaced in three years (see methods).

the reservoir area according to the reservoir morphometry and flooded area. CO_2 concentration is an important component to the gross $CO_{2\text{-EQ.}}$ emissions, in both young and older reservoirs. Methane, on the other hand, has less importance as the reservoir ages (Figure 14). Analysis performed by Jean P. Ometto et al. ("*Patterns of carbon emission as a function of energy generation in hydroelectric reservoirs*" in review) suggested a trend of decreasing $CO_{2\text{-EQ.}}$ emissions with reservoir aging (or with time after impoundment) and a different perspective of the emissions when associated to the amount of energy produced by the system.

For instance, Serra da Mesa reservoir has a high GHG emission profile, however this is a high-energy producing system. The hydropower emissions and energy produced by reservoirs, when compared to fossil fuel electricity production, can be categorized as a low emission renewable energy source.

Other interesting aspects can be extracted from Figure 14. In Corumbá reservoir the GHG emissions reflect the anthropogenic load of organic matter from the untreated sewage. The CO_2 emissions in Funil stress the trophic condition of the reservoir.

Manso showed reduction in both CO_2 and CH_4 from the first to the second sampling period, however the reduction in CO_2 emissions was slight more pronounced. Serra da Mesa, had high $CO_{2\text{-EQ.}}$ emissions which reflects its young age. The cascade reservoirs (Furnas, M.Moraes, LCB Carvalho) low $CO_{2\text{-EQ.}}$ emissions reflect the reduction of anthropogenic carbon output up-stream the reservoir.

CONCLUSION

The outcomes from the project "*Carbon Budget in FURNAS Hydroelectric Reservoirs*" provide a substantial data of Brazilians hydropower reservoirs located in the Cerrado biome (data available, under demand, at www.dpi.inpe.br).

We found an important temporal seasonality on carbon input and output in the studied reservoirs, potentially reflecting in the internal biological processes. We also observed a negative relation between CH_4 bubble emissions in the water-air interface and the age of the reservoir, likewise a negative relation between CO_2 diffusive fluxes in the sediment-water interface and reservoir age.

Physical processes are important drivers for GHG emissions in reservoirs, as well as the spatial distribution of pCO_2 and consequent emissions in the lake surface area. The GHG emissions and lake area showed a positive correlation.

Overall, Cerrado reservoirs showed lower GHG emissions in relation to the energy produced when compared to coal or oil thermoelectric energy generation.

ACKNOWLEDGMENTS

We thank laboratory and field technicians from all Institutes and Universities involved in the Project and field. Water chemistry and GHG analyses were performed at the Federal University of Juiz de Fora (UFJF), Federal University of Rio de Janeiro (UFRJ), São Carlos Institute of Ecology (IIEGA). The National Institute for Space Research (INPE) acquired and analyzed the SIMA data. Support for authors were provided by Conselho Nacional de Desenvolvimento Científico e Tecnológico (CNPq) and Fundação de Amparo a Pesquisa do Estado de São Paulo (FAPESP). FURNAS Centrais Elétricas SA provided funds and logistical support. We are grateful to Jim W. Reid (Academic English Associates) for revising the English.

We also want to thanks Ednaldo Oliveira dos Santos, Carlos Henrique, Eça D´Almeida Rocha, Sambasiva Rao Pacthineelan, Elizabeth Sikar, Claudio Roberto Oliveira da Silva, Josiclea Pereira Rogério, Ayr Manoel Bentes Junior, Adilson Elias Xavier, Marcelo Bento Silva, Rodrigo Santos Costa and Corbiniano Silva, Ivan Bergier, Edmar Mazzi, Plinio Alvalá..

REFERENCES

Abe, D. S., Adams, D. D., Galli, C. V. S., Sikar, E. and Tundisi, J. G. (2005). Sediment greenhouse gases (methane and carbon dioxide) in the Lobo-Broa Reservoir, São Paulo State, Brazil: Concentrations and diffuse emission fluxes for carbon budget considerations. *Lakes and Reservoirs: Research and Management* 10:201-209.

Adams, D. D. (1994). Sampling sediment pore water . In A. M. D. MacKnight (Ed.), *Handbook of Techniques for Aquatic Sediments Sampling*, (2nd ed, , pp. 171–202.) CRC Press, Boca Raton, FL, USA.

AEO: (2009), 'Annual Energy Outlook with projections to 2030', International Energy Agency, www.eia.doe.gov/oiaf/aeo/.

Assireu, A. T., Alcântara E., Novo E.M.L.M., Roland F., Pacheco F. S., Stech J. L., Lorenzzetti J. A. (2011). Inside the hydro-physics processes at the plunge point location: an analysis by satellite and in situ data. *Hydrol. Earth Syst. Sci. Discuss.*, 8, 1–31.

Assireu, A. T., Stech, J. L., Marinho, M. M. , Cesar, D. E. , Lorenzzentti,J. A. , Ferreira, R. M. , Pacheco, F. S., Roland, F. (2005). Princípios Físicos e Químicos a Serviço da Limnologia - Um Exercício. In: Fábio Roland, Dionéia E. Cesar, Marcelo Marinho. (Eds.). *Liçoes de Limnologia*. (1 ed v. 1, p. 487-505) São Carlos: Rima Editora,

Cerri, C. C., Maia, S. M. F., Galdos, M. V., Cerri, C. C. E. P., Feigl, B. J., Bernoux, M. (2009). Brazilian greenhouse gas emissions: the importance of agriculture and livestock. *Sci. Agric*. (Piracicaba, Braz.), v.66, n. 6, p.831-843

Chen, Y. J., Wu S. C., Lee B. S., Hung C. C.. (2006) Behavior of storm-induced suspension interflow in subtropical Feitsui Reservoir, Taiwan. *Limnol. Oceanogr*. 51: 1125-1133.

Cole J. J.and Caraco N. F.. (2001). Carbon in catchments: connecting terrestrial carbon losses with aquatic metabolism. *Mar Freshw Res* 52:101–10.

Cole, J. J., Caraco N. F., Kling G. W., Kratz T. K.. (1994). Carbon-Dioxide Supersaturation in the Surface Waters of Lakes. *Science* 265:1568-1570.

Dean, W.E. and Gorham, E. (1998) Magnitude and significance of carbon burial in lakes, reservoirs, and peatlands. *Geology*, 27 (6), p 535-538.

Devol, A. H., Zaret T. M., Forsberg, B. R. (1984). Sedimentary organic matter diagenesis and its relation to the carbon budget of tropical Amazon floodplain lakes. *Int. Ver. Theor. Angew. Limnol. Verh.* 22: 1299-1 304.

Duchemin, E., Lucotte, M., Queiroz, A. G., Canuel R., DaSilva H. C. P. and Almeida D. C. et al., (2000). Greenhouse gases emissions from a 21 years old tropical hydroelectric reservoir, representativity for large scale and long term estimation. *Veranlundgen Int Vereinigung Theor Angew Limnol* 27, p. 1391.

Ford, D. E. (1990). Reservoir Transport Processes. In Thornton, K.W., Kimmel, B.L., Payne, F.E. (Eds). *Reservoir Limnology: Ecological Perspectives*. Wiley-Interscience: New York, pp. 15-41.

Friedlingstein P., Houghton RA, Marland G, Hackler J, Boden TA, Conway TJ, Canadell JG, Raupach MR, Ciais P, Le Quéré C. (2010) Update on CO2 emissions. *Nature Geoscience*, DOI 10.1038/ngeo_1022, Online 21 November.

Galy-Lacaux C., Delmas R., Jambert C., Dumestre J. F., Labroue L., Richard S. and Gosse P. (1997). Gaseous emissions and oxygen consumption in hydroelectric dams: A case study in French Guyana. *Global Biogeochemical Cycles* 11: 471–483.

Giles, J. (2006) Methane quashes green credentials of hydropower. *Nature* 444:524-525.

Hesslein R. H., Rudd J. W. M., Kelly C., Ramlal P., Hallard A. (1991) Carbon dioxide pressure in surface waters of Canadian Lakes. In: Wilhelms S. C., Gulliver J. S. (Eds) *Air–water mass transfer*. American Society of Civil Engineers, Washington, DC, pp 413–431S

IPCC. (2007). *Intergovernmental Panel on Climate Change. Fourth Assessment Report*: www.ipcc.ch.

Kalff, (2001). *Limnology*. (1ª ed.) Benjamin Cummings, 592 pp ISBN-10: 0130337757. ISBN-13: 9780130337757

Kirchman D., K'Nees E., Hodson R. (1985). Leucine incorporation and its potential as a measure ofprotein synthesis by bacteria in natural aquatic systems. *Appl. Environ. Microbiol.*,49,599-607.

Kortelainen P., Pajunen H., Rantakari M., Saarnisto M., (2004). A large carbon pool and small sink in boreal Holocene lake sediments. *Glob Chang Biol* 10, pp. 1648–1653

Kremer, J. N. , Reischauer, A. , D'Avanzo C. (2003) Estuary-specific variation in the air-water gas exchange coefficient for oxygen. *Estuaries* 26:829-836

La Rovere E. L., Romeiro A.R. (2003) *The Development and Climate Project phase I: Country Study Brazil*. Centro Clima, COPPE, Federal University of Rio de Janeiro, Brazil.

Lelieveld, J., P. J. Crutzen, and F. J. Dentener. (1998). Changing concentration, lifetime and climate forcing of atmospheric methane. *Tellus Ser. B*, 50:128-150.

LeQuere, C., Raupach, M., Canadell, J., Marland, G., Bopp, L., Ciasa, P., Conway, T., Doney, S., Feely, R., Foster, P., Friedlingstein, P., Gurney, K.,

Hougton, R., House, J., Hungtingford, C., Levy, P., Lomas, M., Majkut, J., Metzl, N., Ometto, J., Peters, G., Prentice, I. C., Randerson, J., Running, S., Sarmiento, J., Schuter, U., Sitch, S., Takahashi, T., Viovy, N., Werf, G. V. D., Woodward, F. (2009). Trends in the sources and sinks of carbon dioxide. *Nature Geosci.*, 2, 1831–1836, doi:10.1038/ngeo1689,.

Liikanen A, Flöjt L, Martikainen P. (2002). Gas dynamics in eutrophic lake sediments affected by oxygen, nitrate, and sulfate. *J Environ Qual.* Jan-Feb;31(1):338-49.

Lima, L.B.T., Novo, E. M. L. M. , Ballester, M. Y. R., Ometto, J. P. H. B. (1998) Methane production, transport and emission in Amazon hydroelectric plants. In: *IEEE IGARSS'98 International Geoscience and Remote Sensing Symposium*, 1998, Seattle. Proceedings, p. 2529-2531.

Martin J. L., McCutheon S. C. (1999). *Hydrodynamics and transport for water quality modeling*. Lewis.

Matvienko, B., Sikar, E., Rosa, L. P., Santos, M. A., De Filippo, R., Cimbleris, A. C. P. (2000). Gas release from a reservoir in the filling stage. *Proceedings of the International Association of Theoretical and Applied Limnology*, v. 27, p. 1415-1419.

Moreira-Turcq, P., Jouanneau, J. M., Turcq, B., Seyler, P., Weber, O., Guyot, J. L. (2004). Carbon sedimentation at Lago Grande de Curuai, a floodplain lake in the low Amazon region: insights into sedimentation rates. *Palaeogeography, Palaeoclimatology, Palaeoecology*, 214(1-2): 27-40.

Roland, F., L. O. Vidal, F. S. Pacheco, N. O. Barros, A. Assireu, J. P. H. B. Ometto, A. C. P. Cimbleris, and J. J. Cole. (2010). Variability of carbon dioxide flux from tropical (Cerrado) hydroelectric reservoirs. *Aquatic Sciences* 72:283-293.

Roland, F., N. F. Caraco, and J. J. Cole. (1999). Rapid and precise determination of dissolved oxygen by spectrophotometry: Evaluation of interference from color and turbidity. *Limnology and Oceanography* 44:1148-1154.

Rosa, L. P. , Santos, M.A. , Santos, E. O. , Sikar, B. M. , Sikar, E. . Greenhouse Gas Emissions From Hydroelectric Reservoirs in Tropical Regions. *Climate Change*, v. 66, p. 09-21, 2004.

Rosenberg D. M. , Berkes F., Bodaly R. A., Hecky R. E., Kelly C. A., Rudd J. W. M. (1997). Large-scale impacts of hydroelectric development. *Environmental Reviews* 5: 27-54.

Santos, M. A. , Rosa, L. P. , Sikar, B. M. , Santos, E. O. , Rocha, C. H. E.A., , Sikar, E. , Bentes Junior., A. M. P. (2009). Estimate of degassing greenhouse gas emissions of the turbined water at tropical hydorelectric reservoirs. *Verhandlungen - Internationale Vereinigung fur Theoretische und*

Angewandte Limnologie / Proceedings of the International Association of Theoretical an, v. 30, p. 834-837, 2009.

Santos, M. A., L. P. Rosa, B. Sikar, E. Sikar, and E. O. Dos Santos. (2004). Gross greenhouse gas fluxes from hydro-power reservoir compared to thermo-power plants. *Energy Policy* 34:481–488.

Sikar, E., Matvienko, B., Santos, M. A., Rosa, L. P., Silva, M. B., Santos, E. O., Rocha, C. H. E. D., Sentes A. P. Jr. (2009). Tropical reservoirs are bigger carbon sinks than soils. *Verh. Internat. Verein. Limnol.* vol. 30, Part 6, p. 838-840.

Simon M., Azam F., (1989). Protein content and protein synthesis rates of planktonic marine bacteria. *Mar. Ecol. Prog. Ser.*, 5I , 20l - 213.

Smith, D. C., Azam, F. (1992). A simple, economical method for measuring bacterial protein synthesis rates in seawater using 3H-leucine. *Marine Microbial Food Webs* 6: 107-114.

Smith, L. K, Melack, J. M., Hammond, D. E. (2003) Carbon, nitrogen, and phosphorus content and 210 Pb-derived burial rates in sediments of an Amazon floodplain lake. *Amazoniana*, 17, 413–436.

Smith-Morrill, L., (1987). *The exchange of carbon, nitrogen, and phosphorus between the sediments and water-column of an Amazon floodplain lake*. PhD dissertation, University of Maryland, 209 pp.

St. Louis V. L., Kelly C. A., Duchemin E.,. Rudd J. W. M and Rosenberg D. M. (2000). Reservoir surfaces as sources of greenhouse gases to the atmosphere: a global estimate. *Bioscience* 50, pp. 766–775

Stech J. L., Lima I.B.T., Novo E.M.L.M., Silva C.M., Assireu A.T., Lorenzzetti J.A., Carvalho J.C., Barbosa C.C., Rosa R.R. (2006). Telemetric monitoring system for meteorological and limnological data acquisition. *Verh. Internat. Verein. Limnol.* no. 29, pp. 1747-1750.

Tundisi, J. G. (1990). *Distribuição espacial, sequência temporal e ciclo sazonal do fitoplâncton em represas: fatores limitantes e controladores* Rev Bras Biol, Rio de Janeiro, v. 50, n. 4, p. 937-955.

Wanninkholf, R., McGillis, W. R. (1999). A cubic relationship between air-sea CO2 exchange and wind speed. *Geophys. Res. Lett.*, 26, 1889-1892.

Wetzel, R.G., and Likens, G.E.. (1991). *Limnological Analyses.* 2nd. Ed. Springer-Verlag. 391 pp.

WEO: (2007), *World Energy Outlook , International Energy Agency*. ISBN: 978-92-64-02730-5. www.iea.org/weo/2007.asp, p. 600.

In: Energy Resources
Editor: Enner Herenio de Alcantara
ISBN: 978-1-61324-520-0

Chapter 6

EFFICIENCY OF FOREST CHIP TRANSPORTATION FROM RUSSIAN KARELIA TO FINLAND

Vadim Goltsev*[*]*, Maxim Trishkin and Timo Tolonen
The Finnish Forest Research Institute, Joensuu, Finland

ABSTRACT

The development of modern cut-to-length harvesting techniques, available wood resources and Russian customs policy have created opportunities to export forest chips from Russian Karelia to Finland. An important factor for the export is the costs of transportation of the forest chips. In this study the efficiency and costs of cross-border transportation of forest chips were analysed and compared with the efficiency and costs of transportation of forest chips of Finnish origin.

Data collected from various companies involved in forest chip production and their supply from Russia to Finland were used to calculate the costs of cross-border transportation of forest chips and to estimate the average productivity of chip trucks delivering from Russia to Finland. These outputs were compared with the transportation costs and productivity of chip trucks within Finland and Russia. Truck drivers involved in cross-border transportation of forest chips were also interviewed to determine factors

---

[*]E-mail: vadim.goltsev@metla.fi; Tel: +358 408 015 274.

affecting the efficiency of forest chips transportation. In addition, the quality characteristics of the Russian forest chips being supplied to Finland were analysed.

Analysis of transportation costs showed that the highest costs for the 80 km reference distance are those within Finland – 4.7 €/loose m^3, the costs on the cross-border route studied, from Lendery (Republic of Karelia) to Lieksa (Finland) through the Inari border crossing point, are 3.4 €/loose m^3 and transportation costs within Russia are 3.5 €/loose m^3. Transportation costs as a proportion of the total supply costs were highest for forest chips imported from Russia at 26%, whereas in Finland and Russia they were 23% and 19% respectively. According to the results of the interviews, bad road conditions and idle time on the border were recognized as the main factors decreasing the efficiency of cross-border transportation. Analysis of the quality characteristics of forest chips exported from Russia to Finland did not reveal major differences compared to forest chips of Finnish origin.

Keywords: export of forest chips, quality, productivity, transportation costs.

1. Introduction

The Republic of Karelia is part of the Northwest Federal District of the Russian Federation and represents 1.06% (180 500 km^2) of the country's territory. The western border of Karelia is the state border between the Russian Federation and Finland (Figure 1).

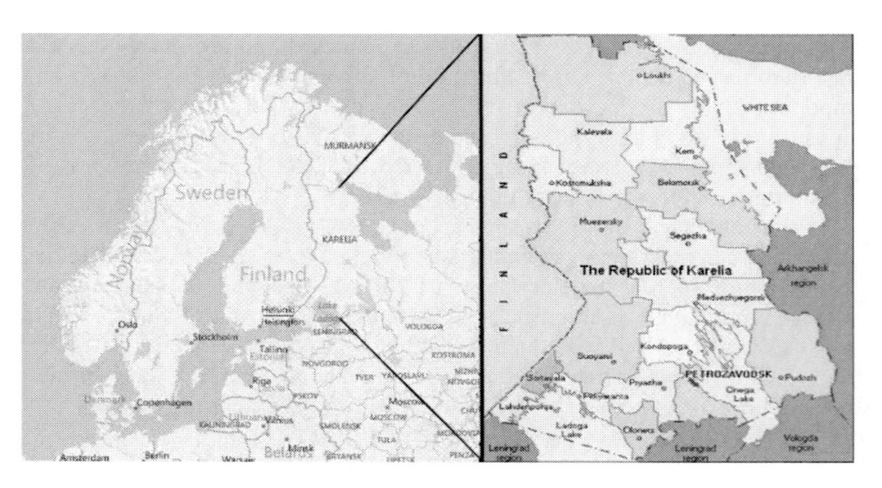

Figure 1. The Republic of Karelia (based on Bing maps).

The population of the Republic of Karelia is about 688 000 inhabitants, with over 75% living in the urbanized areas (Kareliastat 2009). There are three main towns in Karelia: Petrozavodsk (283 000 inhabitants), Kostomuksha (32 500 inhabitants) and Sortavala (20 200 inhabitants). The population density in the Republic of Karelia is only 4 inhabitants per km^2 (The Republic of Karelia in brief 2010). For comparison, in Finland the population density is 16 inhabitants per km^2 (Eurostat 2009).

The total forest area of Karelia is about 14.9 million ha, with a growing stock of 946 million m^3. Forests cover approximately 53% of the territory (Kareliastat 2008). The annual allowable cut is 8.8 million m^3, whereas the total actual cut was about 5.7 million m^3 in 2009. (Ministry of the Forest Complex of the Republic of Karelia 2010) Hence, the utilization rate of annual allowable cut is about 65%, which is the highest in the Russian Federation.

The main harvesting methods used in Karelia are cut-to-length, tree-length and full-tree methods (Syunev et al. 2009). The cut-to-length method is relatively new for Karelia, but its share has grown rapidly and recently it has become the dominant harvesting method. In 2009 the share of the cut-to-length method reached 93% of the total harvested volume, whereas in 2000 it was 42% (Ministry of the Forest Complex of the Republic of Karelia 2010).

The Republic of Karelia has high potential for intensification of fellings and as a result production of forest chips can be also increased. According to Gerasimov and Karjalainen (2009a), the potential volume of energy wood[1] from harvesting in Karelia is 2.3 million m^3, which includes non-industrial roundwood (62%), lifted stumps (18%), unused branches (9%) and defective wood from logging (11%). There are estimations (Regional'naya tselevaya programma... 2007) showing that it is feasible to harvest about 26% of all logging residues, including unused branches, defective wood and non-industrial wood, for energy purposes.

Forest chips could be a source of energy for many communities and industries in Russia. However, the domestic use of bioenergy resources is hindered to some extent by the current policy of expansion of gas pipeline networks to the regions and also by the intensification of energy generation from other renewable sources, mainly hydro energy (Energeticheskaya strategiya… 2009).

At the same time, regions which are not connected to the natural gas grid are dependent on highly priced fossil fuels. The long transportation distance is the major factor that dramatically increases the total cost of fossil fuels in Russia (OECD/IEA 2003). Many fossil-fuel-deficient regions face frequent shortages of

[1]Energy wood – woody biomass used for production of wood-based fuels.

fuel supplies due to weather and transportation conditions and suppliers' preference for exporting fossil fuels. But now the situation is gradually changing. The government of the Republic of Karelia launched two programmes aiming to increase the proportion of locally produced fuels (e.g. firewood, forest chips and peat) in energy production and decrease the dependence on fossil fuels (Regional'naya celevaya programma... 2007, Regional'naya strategiya razvitiya... 2010).

In Karelia, woody biomass is a relatively new fuel in larger scale municipal and industrial energy production, but in the form of firewood it is a common energy source for households, especially in rural areas.

Besides private households, forest industry companies and municipal heat plants are the main users of woody biomass in Karelia (Raitila et al. 2009, Gerasimov and Karjalainen 2009b). Usually, the forestry companies work together with municipalities and supply wood fuel to the municipal power plants. The pulp and paper industry has about 30 woody biomass steam boilers in Karelia (Raitila et al. 2009). Existing biofuel power plants use mainly forest chips and sawmill residues.

However, sawdust has been increasingly used for pellet production in Northwest Russia and because there has been very little demand on the local market for advanced wood fuels, the Russian biofuel industry has so far been mostly export-oriented (OECD/IEA 2003). However, domestic consumption of pellets in Russia is growing (Rakitova et al. 2009).

The use of woody biomass for energy production in Karelia contributes 10% of total energy supply and most of the energy wood is combusted for heat generation (Grigoryev 2007).

However, at the same time about 54% of all the heat plants in the Republic of Karelia use, at least in part, local biofuels, including firewood, forest chips and peat (Regional'naya celevaya programma... 2007, Regional'naya strategiya razvitiya... 2010). In some districts of the Republic of Karelia energy wood is used more widely – in Kostomukshsky, Muezersky and Kalevalsky districts firewood consumption is about 23% of primary energy consumption (Raitila et al. 2009).

For the whole Russian Federation the use of energy wood is much lower and represents only 3% of the total energy generation. The relatively low utilization rate of wood biomass for energy purposes is caused by uncertainties concerning the costs of supply and its availability at a reasonable price level (Gerasimov and Karjalainen 2009b).

In addition to the increased use of wooden biomass locally, there is potential to export forest chips from Russian Karelia to Finland. The current weak local demand for low-quality roundwood from final fellings and thinnings in Karelia makes large volumes of raw materials available for chipping at a reasonable cost, while in neighbouring Finland the demand for forest chips and their utilization is high. In addition, customs duties for forest chips are lower than for other assortments, e.g. saw logs and pulp wood, accounting for only 5% of their export value (Federal Customs Service of Russia 2009).

Figure 2. shows the dynamics of forest chip export from the Republic of Karelia to Finland.

As it can be seen, the export of forest chips increased substantially in 2009. This may be explained by the high customs duties set for pulp wood. Besides that, the demand for energy wood has also increased in Finland resulting in competitive prices of forest chips. Table 1 shows the average price of forest chips (Free Carrier contract agreement) in the Republic of Karelia (MED 2011) in 2010 and the average cost of forest chips (as received) paid by Finnish power plants (PÖYRY 2010) in summer 2010.

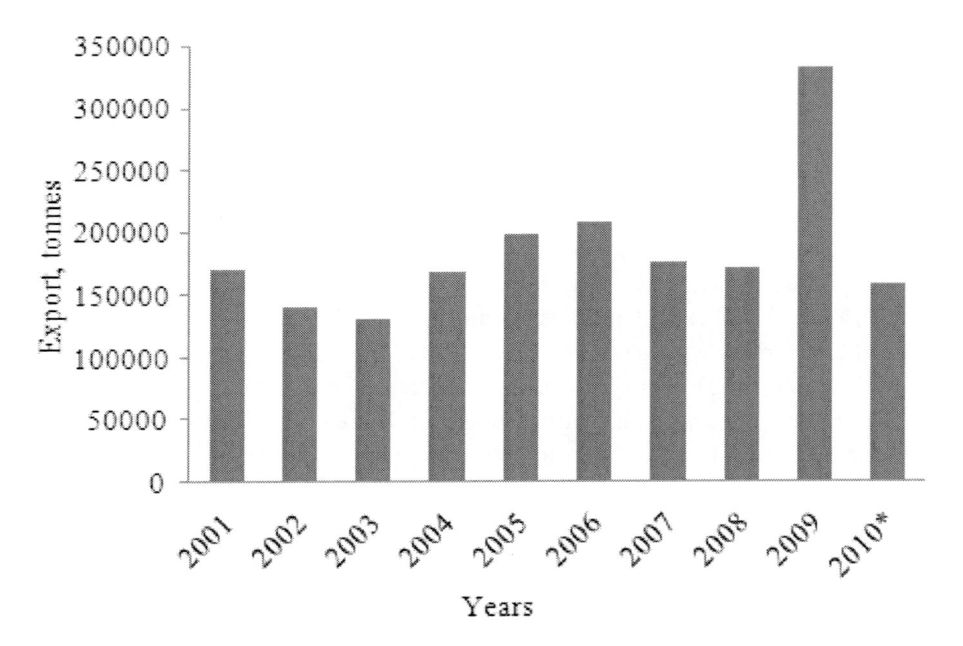

*01.01.2010 – 01.06.2010 (only for first 5 months of 2010)

Figure 2. Export of coniferous forest chips from the Republic of Karelia to Finland (Kareliyastat 2010).

Table 1 The average price of forest chips in the Republic of Karelia and the average cost of forest chips paid by Finnish power plants

Fuel	Producer's price in Russia, €/MW h	Cost paid by Finnish power plant, €/MW h
Pine forest chips	9.70	17.9-18.2
Birch forest chips	11.00	

As it can be seen from Table 1 there is a big difference between the producer's price of forest chips in the Republic of Karelia and the average cost of forest chips paid by Finnish power plants.

Relatively poor preconditions for domestic utilization of forest chips and the significant resource potential create good opportunities for the export of forest chips from Karelia to Finland, where the current energy and climate strategies support the use of forest chips for energy generation. (Renewable energy policy review 2009, Ministry of Trade and Industry 2000). In the past, use of forest chips on a large scale has been less common in Finland (Ranta 2005) and has traditionally been locally-oriented (Heinimo 2008). The forest chips boom started in Finland at the end of 1990s, when the Wood Energy Technology Program 1999-2003, aimed at the development and commercialisation of the use of forest chips, was launched as one of the Government's tools for implementation of the Action Plan for Renewable Energy Sources (Hakkila 2004). As a result of this program, the competitiveness of chips as a fuel significantly improved. The use of forest chips has grown from 1.7 million m^3 in 2001 (Hakkila 2004) to 6.1 million m^3 in 2009, and it is expected that 13.5 million m^3 of forest chips will be used in 2020 (Puun energiakäyttö 2010).

In Finland, forest chips for energy purposes are mainly produced from logging residues, and pruned and unpruned small-size trees (Hakkila 2004), and only a minor part is made of large size roundwood, stumps and roots. On the contrary, in the Republic of Karelia, due to the local features, a better raw material for chipping is roundwood, which is widely available at reasonable cost. Therefore, in Karelia logging residues and small-size trees, raw materials with low bulk density, are not used for chipping. Due to the higher bulk density of roundwood, its harvesting, transportation and chipping are more efficient than that of uncompacted logging residues and small-size trees. Besides, the quality characteristics of chips produced from logging residues or roundwood may differ. For example, forest chips produced from logging residues can be contaminated with soil and stones.

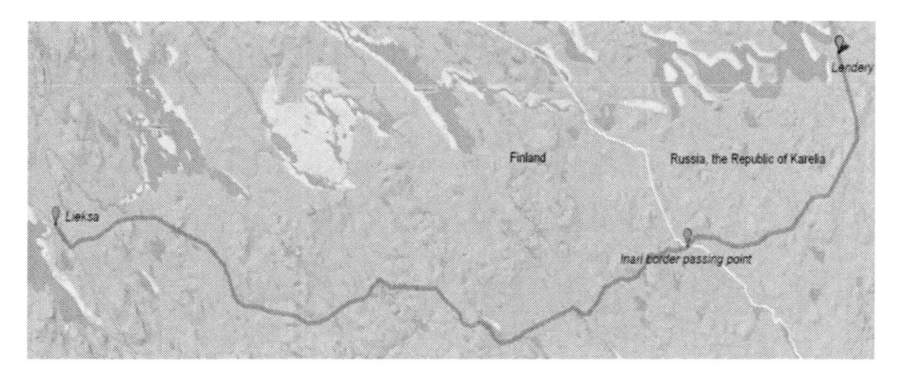

Figure 3. Forest chip transportation route from the Lendery terminal in Russia to the Lieksa power plants in Finland (based on Google maps).

The use of different raw materials for chipping makes it very difficult to compare straightforward overall productivity of the Finnish and Russian forest chip supply systems.

Therefore, this study focuses on the efficiency of the transportation stage, a parameter which can be compared between the countries irrespective of the raw material used for chipping. The study analyses the efficiency of forest chip transportation by trucks from Russian Karelia to Finland, compared with forest chip transportation within Finland. In addition, the quality of forest chips supplied from Russia to Finland is analysed.

The transportation route for forest chips can be seen in Figure 3, beginning from the Lendery terminal in Russia and ending at the power plant in Finland. The distance from the terminal to the border is 25 km and from the border to the power plant in Lieksa is 57 km, making a total of 82 km.

Specific tasks of the study are:

- Analysis of the efficiency of forest chip transportation by chip trucks from Russian Karelia to Finland based on results of the case study on the forest chip supply from the Lendery wood terminal in the Republic of Karelia to the power plants in Lieksa in Finland;
- Comparison of the efficiency of cross-border forest chip transportation from Russia to Finland with transportation efficiency within Finland in terms of costs and transported volumes;
- Measurement of moisture, calorific value, ash content and particle size variation of forest chips transported from the Republic of Karelia to Finland;

- Interviewing of truck drivers to obtain subjective descriptions of factors affecting the efficiency of forest chip transportation and to identify measures to improve it.

2. Materials and Methods

2.1. Analysis of Forest Chip Transportation Efficiency

The study is focused only on the cross-border transportation of forest chips by trucks. Other types of forest chip transport are not considered here. Cross-border transportation of forest chips is the final stage of the Russian-Finnish forest chip supply chain, which begins in this case from felling sites in the Republic of Karelia and ends at power plants in Finland. In the case study felling, delimbing and cross-cutting is done by harvester, based on the caterpillar excavator Fiat-Kobelco E135SR with the installed harvesting head Kesla 22RH (Figure 4). The haulage of assortments to the roadside is carried by the forwarder Timberjack 1010D (Figure 4) with a payload capacity of about 15 m^3. Then logs designated for chipping are transported by Volvo log trucks from the roadside, separately from industrial wood, to the terminal in Lendery village. Comminution is done at the terminal as a reasonable compromise between chipping at a landing and at a power plant in Finland, because Russian export duties for roundwood, including energy wood, are higher than export duties for forest chips (Federal Customs Service of Russia 2009). At the same time, for the chip producers it is important that forest chips from Russia to Finland have higher added value than unprocessed roundwood for energy purposes produced in Finland.

Figure 4. Excavator-based harvester Fiat-Kobelco E135SR (left side, photo: A. Seliverstrov) and forwarder Timberjack 1010D (right side, photo: V. Katarov).

Figure 5. Mobile chipper Heinola 1310RML working on a terminal in the Republic of Karelia, Russia (photo: Y. Suhanov).

Figure 6. 7-axles (left) and 6-axles (right) chip trucks.

At the Lendery terminal the logs are chipped into cone-shaped piles by a chipper Heinola 1310RML installed on a truck (Figure 5).

The core part of the supply chain is the transportation of forest chips by trucks from the terminal in Karelia to the heat plants in Lieksa, Finland. Loading of the trucks is done at the Lendery terminal by a bucket front-loader. The average loading capacity of the truck is about 100 loose m^3 with trailer and about 40 loose m^3 without trailer. Russian transport norms (Instrukciya po perevozke... 1996) strictly limit the maximum allowable payload on one axle of a chip truck. Therefore, the use of a chip truck with more axles (Figure 6) is more reasonable due to the bigger allowable payload.

On the given transportation route, chip trucks cross the border at a temporary border crossing point located near to the village of Inari. At the crossing point the trucks are weighed and the drivers must present all the documents required by the border authorities, including phytosanitary certificates. The drivers are sometimes also asked to present their trucks for technical inspection. It should be noted that these formalities are not always followed by the border passing point personnel with the same care.

After the border, the trucks drive about 57 km and unload at the Lieksa power plants using a paddle chain system. One of the heat plants in Lieksa is a company-owned boiler house which was initially used only for heat generation for its own sawmill. However, the boiler house is now also used to provide heat for Lieksa municipality. The total annual energy output of this heat plant is 200 000 MWh of primary energy. About 90% of the fuel for energy production comes from its own sources and about 10% comes in the form of forest chips from Russian Karelia.

When examining the cross-border transportation of forest chips, it should be noted that according to the findings from the interviews, the trucks are only fuelled with Finnish diesel, even though the price of Russian diesel is half as much. This is so that the transport company avoids the risks of unplanned repairs caused by low fuel quality.

The volumes presented in this report are given in solid m^3 if not otherwise stated. The conversion factor of 0.40 was used to convert loose m^3 to solid m^3 of wood chips (Hakkila 2004). When necessary, volume units (m^3) were converted into energy units (MWh) or vice versa, assuming that wood has about 50% moisture content and about 2 MWh/m^3 energy content or 0.77 MWh of energy content per 1 loose m^3. The currency exchange rate of the Central Bank of the Russian Federation at 10 September 2010 was used when it was necessary to convert costs in Roubles to Euros. In the conversion, 39.18 Roubles correspond to 1 Euro.

The estimation of cross-border forest chip transportation efficiency is based on analysis of the following information describing the transportation route from the Lendery terminal (Russia, the Republic of Karelia) to Lieksa (Finland, the Province of North Karelia):

- Run parameters of the studied transportation route
- Transportation distance from the Lendery terminal to the Inari border crossing point and from there to the unloading point in Finland for each studied delivery.
- Average driving speed in Russia from the terminal to the Inari border crossing point and average driving speed in Finland from the border to the unloading point for each studied delivery.
- Average duration of loading/unloading operations

- Idle time of chip trucks which consists of breaks longer than 15 minutes – time of loading/unloading, time needed to pass all the border formalities at the Inari border crossing point (including time waiting in a queue) and duration of maintenance and repairs.
- Transportation costs and their shares in overall supply costs in Russia, Finland and for cross-border supply.
- Payloads of a forest chip truck in the cross-border transportation

Data regarding transportation distance, average driving time, loading/unloading operations and idle time were obtained from the tachograph recording system installed in the cabin of the truck and covered a total of 6 runs Lendery-Lieksa-Lendery. The 6 deliveries investigated were made in January and February 2010.

Transportation costs correlating with transportation distance are important factors affecting the import of forest chips from Russia to Finland. According to Ranta (2005), in Finland 100 km is considered the maximum economically acceptable transportation distance. Thus, in the case study, 100 km was chosen as the maximum distance for forest chip transportation and it was used for further comparison. Comparative analysis of the transportation costs includes the Russian-Finnish cross-border route, transportation within Finland and within Russia.

One of the aims of the study was to compare the costs of cross-border transportation of forest chips with the costs of forest chip transportation within Finland and Russia. The costs of cross-border transportation were calculated based on the data collected within the study from a transport company delivering forest chips from Russian Karelia to Finland. The data collected are for 2010 and include costs of fuel, labour, service and insurance. Annual payoff, overhead costs and amortisation costs were calculated according to Gerasimov et al. (2009b). The costs of cross-border transportation do not include value-added tax (VAT) on fuel because companies working on cross-border transportation outside the EU are not obliged to pay VAT (Palvelujen ulkomaankaupan arvonlisäverotus 2010). For comparison purposes, the costs of forest chip transportation within Russia and Finland were taken from the literature: the costs within Finland were valid for year 2003 (Ranta and Rinne 2006) and the theoretical costs calculated by Ilavsky et al. (2007) for Tihvin district of the Leningrad region were valid for 2006. The Finnish costs were indexed to the cost level of year 2010 using the 3% average annual increment of transportation costs in Finland (Tilastokeskus 2010).

Table 2 Annual growth of cargo transportation tariffs in the Leningrad region (Federal Service 2010)

Year	2002	2003	2004	2005	2006	2007	2008	2009	2010
Cost increment, %	18.6	13.5	61.3	17.3	13.17	9.8	30.13	-4.52	19.91*

Notes: *- expected growth calculated based on annual growth during 2002-2009.

The Russian transportation costs were indexed to the cost level of year 2010 using the data on annual growth of cargo transportation tariffs in the Leningrad region of Russia (Federal Service... 2010), as presented in Table 2. The costs of forest chip transportation in the Leningrad region were used for the comparison because no reliable data on costs of forest chip transportation in the Republic of Karelia were available. Also, it is appropriate to apply this data to the Republic of Karelia because of similarities in road conditions.

The transportation costs were compared as €/MWh at 10 km intervals within a 100 km distance. In order to determine transportation costs at 10 km intervals, the linear interpolation method was used for the theoretically calculated costs in the Tikhvinsky and Boksitogorsky districts of Leningrad region, because the published costs referred only to the 20, 60 and 100 km intervals.

The average delivered payload of the chip trucks involved in cross-border transportation was obtained from the company receiving the chips from Russia at its power plants in Finland. Further, this value was compared with the average delivered payload of chip trucks transporting forest chips within Finland. In addition, data on the forest chip flow from one terminal in Russia to Finland was obtained from a company engaged in cross-border transportation of forest chips.

2.2. Laboratory Analysis of the Quality of Forest Chips Supplied from Russia to Finland

The fact that in Finland forest chips are mainly produced from logging residues and small-size trees and in Russia only roundwood is used for production of forest chips makes it interesting to compare the quality characteristics of forest chips of different origins. Sampling of the forest chips coincided with a shortage of forest chips at the Lendery terminal due to an interruption in chipping. Because of the study's time limits, it was not possible to wait until chipping at the Lendery terminal began again. For these reasons, samples of the forest chips were taken from two terminals in Värtsilä and Lahdenpohja (Figure 7 and Figure 8) located in the southern part of Russian Karelia. These terminals use material for chipping that is similar to the Lendery terminal – a mix of coniferous and deciduous roundwood supplied from forests of the Republic of Karelia. Therefore the difference in the quality of forest chips between these terminals should be negligible. Their quality parameters were analysed and compared with the parameters of forest chips produced from logging residues and small-dimension trees in Finland. Both terminals in Karelia use roundwood as raw material for forest chip production and in both cases it includes a mixture of spruce, pine, birch and aspen. The forest chips samples from the terminals were analysed to compare the following quality characteristics of Russian and Finnish forest chips:

- moisture content
- calorific value
- ash content
- particle size distribution

The samples of Russian forest chips dedicated for supply to Finland were obtained in June 2010 according to recommendations given in the standard CEN/TS 14778-1. According to the methodology given in the standard, the stock pile of forest chips in both cases was visually divided into three horizontal layers: upper, middle and bottom. The number of samples taken from each layer was in proportion with the volume contained in each layer. Thus, the number of increments in the pile layers for Värtsilä terminal was 2, 3 and 6 and for Lahdenpohja terminal 1, 3 and 5 accordingly. The samples were taken manually by standard-sized shovel systematically (Alakangas 2005) and equally spaced around the circumference of the heap (CEN/TS 14778-1). Taking samples from the very top of the heap and the bottom 300 mm was avoided. Forest chips to be taken as samples were sorted according to the recommendations made by Alakangas (2005) in order to avoid inclusion of visually bigger pieces of forest chips in the samples.

The samples obtained at each terminal were immediately placed in 7 litre transparent plastic bags (Figure 9) and isolated from outside conditions to maintain stable moisture content. Each bag was labelled with identification of location, position in the pile and serial number (CEN/TS 14778-1). After delivery to the laboratory of the Finnish Forest Research Institute, the samples were kept out of direct sunlight and stored below 5 °C in order to decrease biological activity before laboratory tests (Figure 10).

Figure 7. Forest chips pile at the Värtsilä terminal.

Figure 8. Forest chips pile at the Lahdenpohja terminal.

Figure 9. Taking samples at the terminal.

Figure 10. Packed samples for laboratory analysis.

The samples were obtained, prepared and stored according to CEN/TS 14778-1 and CEN/TS 14780 for determination of moisture content, ash content, calorific value and particle size distribution, based on CEN/TS 14774, CEN/TS 14775, CEN/TS 14918 and CEN/TS 15149 respectively. Moisture content determination required the samples to be dried at 105 °C in air until a constant mass was achieved. The percentage of moisture was then calculated from the mass difference of the samples. Ash content was determined by calculating the mass of the remaining residue after the sample was heated at 550 °C in air under time-controlled conditions. Determination of calorific value included pulverizing of samples by Retsch-1 mill with 10 and 0.5 mm bottom sieves. Then samples were pelletized to 14 mm in diameter with the recommended sample size ranging from 0.8-1.3 grams. Prepared samples were placed on the tared crucible of a sample pan for weighing and were entered into the memory of a calorimeter. The crucible was placed in a sample holder of a bomb and a fuse was attached. The sample holder was placed in the combustion chamber and closed with the bomb cap. The bomb was pressurized up to 3 MPa and placed in the vessel compartment where it was automatically surrounded with water of known volume. The sample was ignited and water temperature was recorded by the calorimeter. Based on the temperature profile, the calorimeter calculated the heating value. For particle size determination and distribution an oscillating screen with apertures of 3.15, 6.3 and 20 mm was used (CEN/TS 15149-1).

2.3. Interviews with Forest Chip Truck Drivers

The interviews with truck drivers had the aim of identifying the impact of different factors on the productivity of their trucks and consequently on the efficiency of forest chip transportation from Russian Karelia to Finland. The drivers' views are important in the interpretation of the current situation regarding transportation efficiency and in finding possible solutions to improve the situation.

The interviews were based on a questionnaire designed to obtain individual responses. It was written in Russian and Finnish versions (Goltsev et al. 2011) because it was planned to interview drivers from both countries. The respondents were interviewed by direct questioning, via post and by the phone.

The questionnaire started with a description of the main goal and purpose of the study in order to make the potential respondents aware of the importance and significance of the topic. There were 27 questions in total: 5 regarding personal background (i.e. education, experience, etc.) and 22 forming the specific part related to forest chip transportation and technical peculiarities.

In every multiple-choice question, only one answer was possible. The 22 questions of the specific part were in a logical order and were to be filled in sequentially to build up a comprehensive picture of the respondent's views. Truthful answers were needed because the reliability of the results is an important issue to rank the main factors influencing transportation efficiency. In total, 11 respondents were interviewed from 4 different companies in both Russia and Finland. Drivers of both nationalities were almsot equally represented, with 5 respondents from Finland and 6 respondents from Russia.

3. Results

3.1. Efficiency of Forest Chip Transportation

The analysis of forest chip transportation efficiency is based on data which describe the qualitative characteristics of the route from the Lendery terminal in Russia to the Lieksa power plant in Finland, including distance, duration of one run, average speed, etc. These data were obtained from the tachograph recording system installed on a chip truck of one of the transporting companies specialising in cross-border transportation of forest chips. Quantitative characteristics of the route, such as average payload of a truck and total volume of forest chips flow, were obtained from the companies supplying chips from Russia to Finland. Table 3 shows the main parameters for the specific route.

Table 3. Observed parameters of the transportation route from the Lendery terminal, Russia, to the Lieksa power plant, Finland

Parameter	Value
One way run, km	82
Average duration of one run, hours	6:38
Average payload, m^3	100
Minimum-average-maximum volume of forest chip flow over 8 months, m^3/month	327-2770-5217
Total volume of forest chip flow for 8 months, m^3	22157

The average transportation distance from the Lendery terminal to the Inari border crossing point was 25 km and from Inari to the unloading point at the Lieksa district heating power plant was 57 km, making a total distance of 82 km. This is less than the 100 km maximum transportation limit defined by Ranta (2002) as cost-competitive in Finnish conditions. For comparison, in Russia forest chips are cost-competitive only if the transportation distance is less than 50 km (Goltsev et al. 2010). This is a good illustration of the difference between costs of fossil fuels and forest chips in Russia and Finland.

The average driving speeds from the Lendery terminal to the Inari border crossing point and from Inari to the Lieksa power plant are presented in Table 4. Table 4. shows that the average driving speed on the Russian side between Lendery and Inari is 34 km/h, whereas on the Finnish side between Inari and Lieksa it is 66 km/h. The average driving speed on the Finnish side is almost twice as high as on the Russian side. The difference between the maximum and the minimum driving speeds is 13 km/h on the Russian side, where the maximum and minimum driving speeds are 43 and 30 km/h respectively. The difference between maximum and minimum driving speeds is 19 km/h on the Finnish side, where the maximum and minimum driving speeds are 76 and 57 km/h respectively. The average driving time is almost the same for the Russian and Finnish parts of the route, despite the big differences between the transportation distances in Russia and Finland. These figures clearly reflect the difference between driving conditions on Russian and Finnish roads. Time taken to load in Lendery and unload in Lieksa, the duration of the run on different sections of the route, and idle time consisting of breaks longer than 15 minutes are shown in Table 5. The time consumption was based on the time records provided by an entrepreneur for 6 runs, amounting to 39 hours 45 minutes of working time altogether. Due to the short period of observations, time spent on maintenance and repair is not shown in Table 5. Table 4 shows that the average time taken by driving in Russia and Finland was 45 and 52 minutes respectively, although the distance on the Finnish side is twice as long as on the Russian side. Crossing the border on average takes 36 minutes, although the maximum and the minimum time spent at the crossing point were 105 and 5 minutes respectively. Loading at the Lendery terminal was normally done at the end of the truck driver's working shift and the duration of this operation was therefore not recorded by the tachograph system on the 6 observed runs. The average duration of the loading operation was estimated based on the interviews with the transport companies, but it was not possible to estimate the maximum and the minimum duration of loading there. Unloading at the Lieksa power plant on average took 50 minutes, with a maximum unloading time of 60 minutes and a minimum of 30 minutes. Lunch breaks on average took 43 minutes, while the minimum and the maximum were 30 and 60 minutes respectively.

Table 4. Speed of chip trucks on the transportation route from the Lendery terminal in Russia to the Lieksa power plant in Finland

Route section	Distance, km	Speed, km/h			Average driving time, h
		Average	Maximum	Minimum	
Lendery ↔ Inari	25	34	43	30	0:42
Inari ↔ Lieksa	57	66	76	57	0:51

Table 5. Duration of main operations on recorded routes

Duration (min)	Average	Maximum	Minimum
Time consumption for driving			
Lendery ↔ Inari	45	50	35
Inari ↔ Lieksa	52	60	45
Time consumption for loading and unloading operations			
Loading (Lendery)*	60	-	-
Unloading (Lieksa)	50	60	30
Idle time			
Border (Inari)	36	105	5
Lunch breaks	43	60	30

Notes: *Only average loading time was given by the entrepreneur.

Figure 11 provides average time distribution within one run for each operation in percentages of the total time, estimated based on the time records from the 6 observed runs. The total duration of the 6 recorded runs was 2385 minutes and the average duration of one run was 398 minutes or 6 hours 38 minutes.

Driving takes only 48% of the total time of one run, due to the relatively short transportation distance. About 22% of the total run time was spent driving the 25 km on the Russian side, and this was almost the same time needed to drive the 57 km on the Finnish side. The next most time-consuming operation is crossing the border, which takes 18% of the total run time. Loading and unloading operations represent relatively small proportions of the total, 15% and 13% respectively. The smallest proportion, 5% of the total run time, was spent on lunch breaks.

The costs of cross-border transportation of forest chips were compared with the transportation costs within Finland and Russia (Figure 12).

The transportation distance on the cross-border route was 82 km and therefore, for the cost comparison, 80 km was used as a reference distance. According to Figure 12, the transportation costs for cross-border transportation are 3.4 €/loose m^3 or 8.5 €/solid m^3. The total supply costs in the case of cross-border transportation of forest chips is 28.8 €/m^3, which comprises 10.95 €/m^3 for harvesting and forwarding, 3.7 €/m^3 for transportation of logs to the terminal (average transportation distance about 50 km), 6.25 €/m^3 for chipping and 7.93 €/m^3 for other costs. All the expenses are given per solid m^3. Road transportation costs for the studied route are 26% of the total supply costs.

The productivity of trucks delivering forest chips from Russia to Finland is an important factor affecting transportation efficiency. Based on data provided by the company, the average load of the trucks is 90-110 loose m^3 coming from Russian Karelia and 120-140 loose m^3 coming from Finland to the power plant. The difference is due to the legislative limitations set in Finland and Russia, where there are differences in the maximum allowable weight of loaded trucks. In Finland, the maximum allowable weight of a truck with 6 axles is 53 tonnes, and for a truck with 7 or more axles it is 60 tonnes (Asetus ajoneuvojen käytöstä.. 1997). In Russian Karelia, the maximum allowable weight of a loaded truck on roads of republic and federal supervision without special permission is 38 tonnes, while with special permission issued by the Republic's road authority FGU Uprdor Kola the maximum allowable weight increases on republic roads to 55 tonnes and on federal roads to 44 tonnes (Instrukciya po perevozke.. 1996). In the case study, on the route from Lendery to Inari the maximum allowable weight can be increased from 38 to 55 tonnes and according to the interviews, drivers receive regularly such permission.

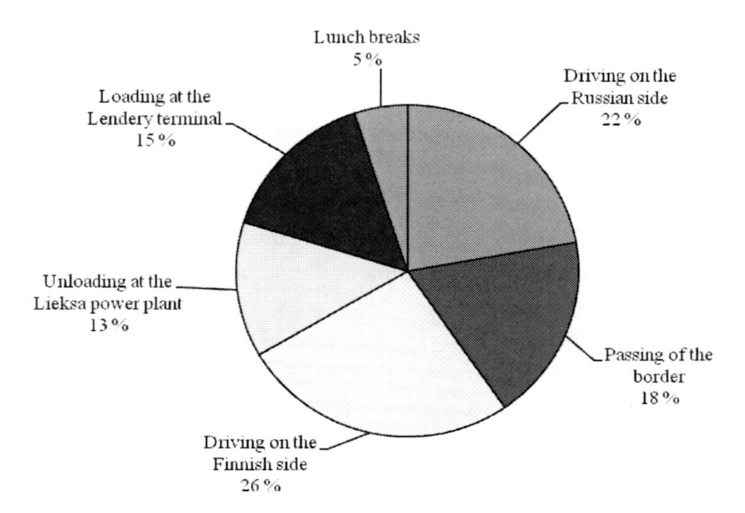

Figure 11. Time distribution on transportation route (based on 6 recorded runs with total time 2385 min).

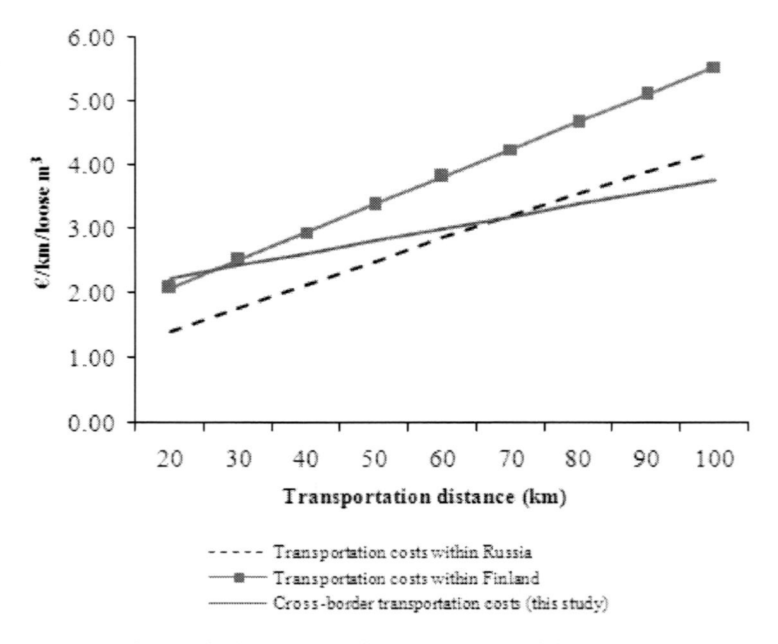

Figure 12. Comparative analysis of costs of cross-border and domestic transportation of forest chips.

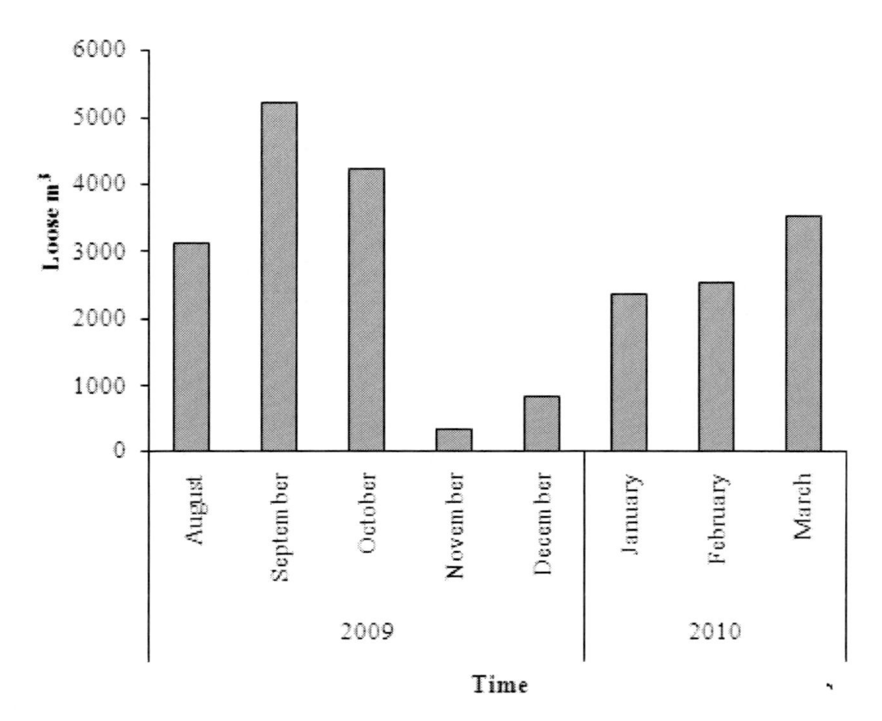

Figure 13.Volumes of forest chips transported throughout the year by the entrepreneur.

Efficiency of supply is also provided by continuity of forest chip flow. The transport company provided data (Figure 13) on the amounts of forest chip supplied throughout the year from Russia to the Lieksa power plant.

As can be seen in Figure 13, the lowest volumes of forest chips were delivered by the entrepreneur in November and December 2009. This was due to the breakdown of the chipper at the Lendery terminal.

Inability to quickly change the source of forest chips made this supply chain vulnerable. This is dangerous especially in winter, when the demand for heat is at its highest.

3.2. Quality of Forest Chips Supplied from Russia to Finland

Laboratory tests have been done to determine the quality parameters of forest chips from the Lahdenpohja and Värtsilä terminals in Russia and to compare them with the official standard CEN/TS 14961. The parameters are compared in Table 6.

Table 6. Quality parameters of the forest chips

Compared parameters	Lahdenpohja, forest chips	Värtsila, forest chips	Standard CEN/TS 14961
Moisture content (%)	47	44	M50 (Lahdenponja), M45 (Värtsilä)
Net calorific value (dry matter) MJ/kg	18.7	18.4	-
Net calorific value of fuel as received MJ/kg	8.8	9.2	-
Particle size distribution (%): >20mm; 20-6.3; 6.3-3.15; <3.15	6, 66, 20, 8	7, 70, 11, 10	P45B (Lahdenpohja, Värtsilä)
Ash content (%)	0.5	1.2	A0.5 (Lahdenpohja), A1.5 (Värtsilä)

Based on the analysis shown in Table 5, the moisture content of the samples is in accordance to the CEN/TS 14961 standard. Forest chips from Värtsilä have 44% moisture content and fulfil the M45 moisture norm and forest chips from Lahdenpohja have 47% moisture content, fulfilling the M50 moisture norm. The net calorific value for the forest chips from Lahdenpohja is 18.7 MJ/kg and from Värtsilä 18.4 MJ/kg, but neither a net calorific value nor a net calorific value as received is regulated by CEN/TS 14961. The particle size distribution of samples from both locations conforms with the P45B size class of the standard. According to the CEN/TS 14961 standard, the ash content for Lahdenpohja is A0.5 class and for Värtsilä it is A1.5 class.

3.3. Opinions of Forest Chip Truck Drivers

The results obtained regarding the respondents' working experience in transportation of forest chips, transportation in general and education indicated high proficiency and long experience in transportation of forest chips. Most of them had worked in forest transportation for more than 5 years, besides which all had more than 10 years of working experience in general transportation. In addition, most of them had completed special education and supplementary courses related to forestry and transportation. Therefore reliable answers to the questions in the specific part of the questionnaire could be expected.

The question regarding the design of trucks used for transportation of forest chips indicated that 50% of the respondents use specially designed trucks, 10% use modified trucks originally designed for other purposes and 40% did not answer. The drivers saw the importance of personal skills as very strong (18%), strong (28%), moderate (45%) and very low (9%) for the transportation productivity of forest chips. A focus on the achievement of the maximum possible productivity was very important for 45% of the drivers, important for 19%, and 36% of the respondents did not answer this question. The interviews did not show clearly how important the achievement of the maximum possible productivity is for the drivers because a relatively large number of respondents did not answer the question. The next question was about the importance of salary as motivation for the drivers to improve their productivity. Salary was a very important motivation factor for 36% of the interviewed drivers, important for 36%, moderate for 19% and 9% did not answer the question. These answers are explained by the fact that about 60% of the drivers are on piece-rate wages, less than 20% of the drivers have an hourly based salary and about 20% get a combination of hourly based and piece-rate wages, depending on work flow and other conditions. Although there was no clear answer to the question about the importance of achieving maximum productivity, the answers regarding the payment system clearly illustrate that the respondents are highly motivated to increase their productivity.

The interviews revealed that in the given case, the efficiency of cross-border transportation of forest chips often suffers due to underloading of the trucks. Of the interviewed drivers, 73% had cases of underloading and only 18% had no cases, while 9% did not answer. Almost half of the drivers felt that underloads have a very strong impact on overall productivity, 27% assumed a strong impact and 27% moderate.

The interview showed that 36% of the respondents were forced to carry out unplanned maintenance or to repair their trucks more than 5 times per year, 45% of the interviewed drivers 2-5 times per year, and only 9% dealt with unplanned maintenance just once a year. For those drivers which answered the question this means an increase in workload of up to 10% for 30% of respondents, from 10 to 20% for 50% of them, and from 20 to 30% for 20% of them. The drivers pointed out several mechanisms whose breakage strongly affects productivity.

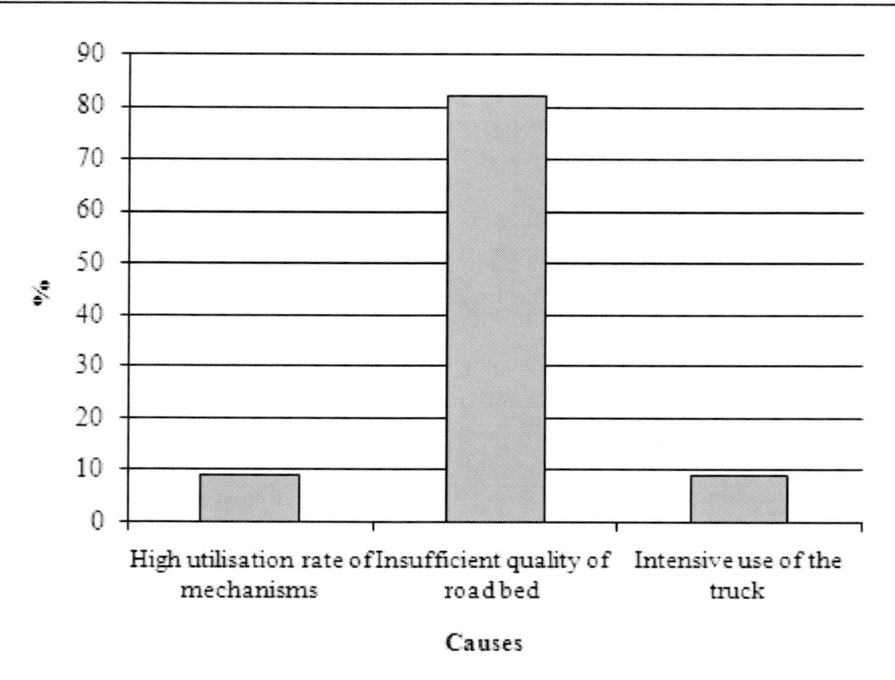

Figure 14. Possible causes of truck breakdown.

About 47% of the respondents mentioned the unloading equipment, as medium maintenance 41% indicated the gearbox, and 12% of the drivers mentioned the fuel supply system and transmission. The results on possible causes of truck breakdown are presented in Figure 14.

As shown in Figure 14, insufficient quality of the road bed is seen as the main cause by 75% of respondents, for 17% high deterioration of base mechanisms is the main factor and the intensive use of the truck is the main reason for just 8%. All the drivers mentioned that the road bed quality and the development of the road network affect transportation productivity very strongly or strongly.

In the drivers' opinion, their productivity depends to some extent on the loading method; 18% consider the relation as very strong, 46% as strong and 18% as moderate or low. All the respondents agreed that a front wheel loader is most efficient for loading.

The relation between the unloading method and the overall productivity is not clear for the drivers: 27% see a strong correlation, 37% thought it moderate, 9% low and 27% very low. The commonly used chain unloading system is recognised as the most efficient.

Regarding the return of deliveries back to Russian Karelia due to bad quality, 36% of the drivers had experienced this while 64% had not. Idle time at the border was recognised by 64% of the drivers as a very important factor affecting overall productivity while 36% considered it an important factor.

The majority of the drivers indicated the idle time during loading/unloading operations as a factor with a moderate influence on the overall productivity of their trucks. Thus, all these factors affect the productivity of cross-border transportation of forest chips. The drivers' estimates of the average actual productivity as percentages of the maximum possible productivity are presented in Figure 15.

Figure 15 shows that, in the drivers' opinion, it is not currently possible to fully utilise the capacities of their trucks. Only 9% estimated that they achieve from 81 to 90% of the maximum possible productivity of their trucks.

Most of the drivers estimated their productivity as between 51% and 80% of the maximum possible. It should be noted that only 9% of the respondents felt their current productivity level was less than 50% of the maximum possible. The drivers pointed out several factors influencing transportation productivity (Figure 16).

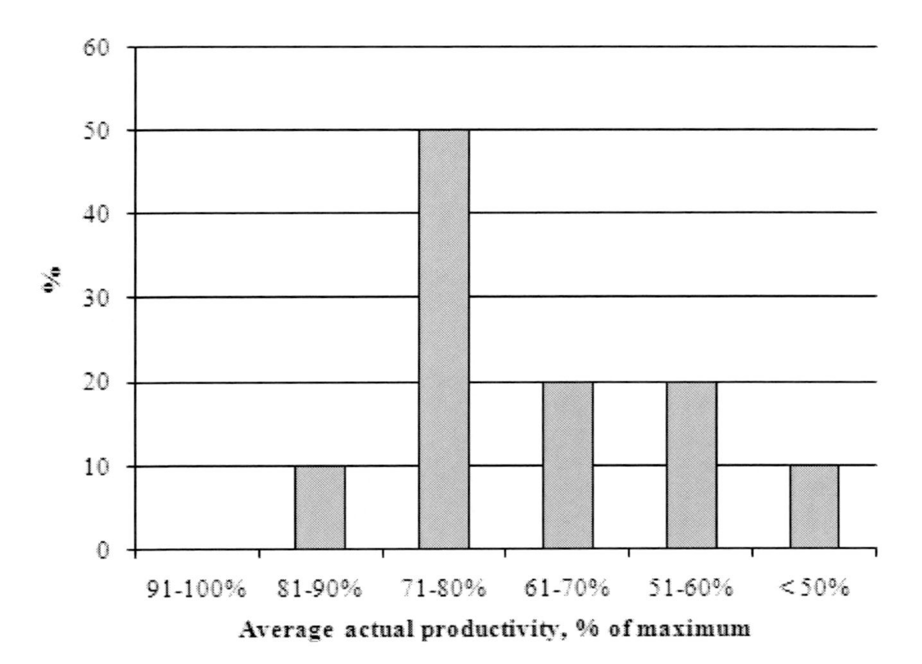

Figure 15. The drivers' estimates of the average actual productivity compared with maximum possible productivity.

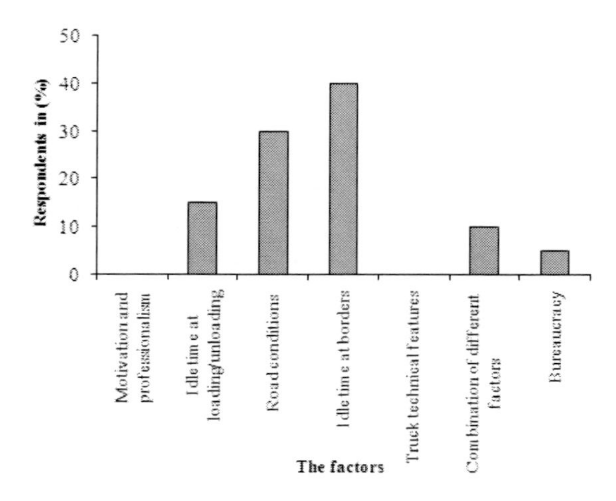

Figure 16. The main factors influencing transportation efficiency.

About 40% of the respondents underlined idle time at the border as the main factor affecting productivity, while 30% of the respondents pointed to road conditions in Russia. Other factors like idle time during loading/unloading operations, bureaucracy at the border crossing and a combination of different factors were named as the main issues affecting productivity by 15%, 10% and 5% of the respondents respectively. The drivers were asked to rank the different factors according to their impact on transportation productivity. The results of ranking are presented in Figure 17.

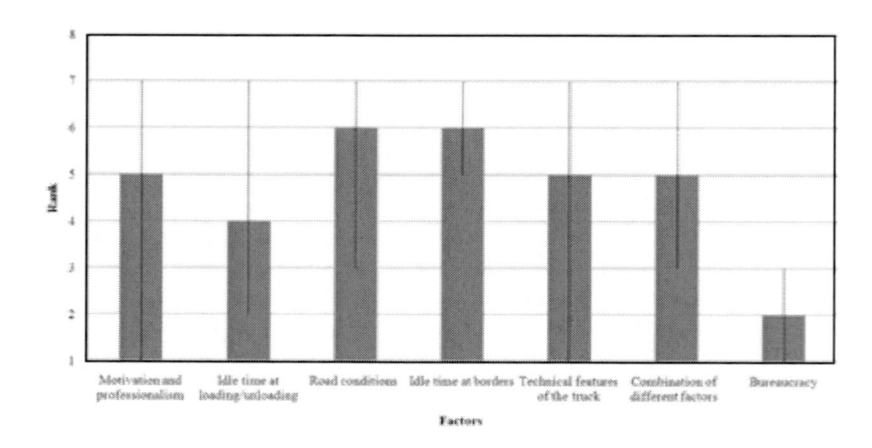

Figure 17. Ranking of the main factors affecting transportation efficiency (1 – least important, 7 – most important; the vertical black line on the each bar indicates the deviation of responses).

As shown in Figure 17, the bad road conditions in Russia and idle time during customs procedures are the most important factors (average ranking 6) affecting transportation efficiency, and those 2 factors also have the smallest range of deviation among the responses. Among the other factors influencing transportation efficiency are technical features of the truck and a combination of different factors (average ranking 5). Motivation and professionalism had an average ranking of 4. It is interesting that the drivers ranked differently the apparently related issues of idle time at the border (average ranking 6) and bureaucracy at the crossing point (average ranking 2).This is because idle time at the border is often caused not by bureaucracy but by the excessively large transport flow at the crossing point.

In addition to these factors, the working schedule of the Inari border crossing point was mentioned as a factor affecting the efficiency of forest chip transportation. This subject was not raised during the interviews but appeared later during the discussion with the director of one of the transport companies working on the cross-border route. The current working schedule at Inari border crossing point is on Mondays 15.00 – 20.00, Tuesday to Thursday 7.00 – 19.00, Fridays 7.00 – 17.00, and on Saturdays and Sundays it is closed. The current working schedule, in the opinion of the transport company, has a negative impact on the transport flow of forest chips and consequently on its efficiency. As a result, in the case being studied the weekly normative plan of 10 round trips to the Lendery terminal was not fulfilled and only 8 round trips at maximum were possible.

On the border between the Republic of Karelia and Finland there are currently three permanent (Lyttä, Värtsila and Suoperä) and twelve temporary border crossing points: Korpiselkä, Ristilahti, Kuolismaa, Haapavaara, Voynica, Inari, Mäkijärvi, Rovkuly, Kolvasjärvi, Kokkojärvi, Kivivaara, Syväoro and Hekselä (Federal Agency for Development...2009). The temporary border crossing points are difficult in terms of their management, some of them are working only for round wood export and, for many reasons, e.g., road conditions and their remote location, they can be out of service for uncertain periods of time. According to the interviews, this has a negative impact on cross-border trading. The two permanent border crossing points (Kostomuksha and Värtsilä) are not able to process the traffic flow if the temporary points are closed.

Another factor affecting the cross-border transportation of forest chips is the duration of the slush seasons in Finland and Russia. During the slush season transportation of wood from forests decreases significantly, which can affect the work of chipping terminals if their reserves of wood are not large enough. Moreover, if a terminal is not connected to a road with a hard surface, transportation of forest chips from the terminal becomes impossible. In the

Republic of Karelia the slush season is one month longer than in Finland. In Finland the slush season normally starts in the third week of April and lasts for one month on average, while in Russian Karelia it starts a month earlier and lasts a month longer. It is the spring slush season, according to the interviews, that has the biggest impact on road accessibility; therefore, the autumn slush season is not taken into account.

CONCLUSION

This study set out to describe and analyse the productivity of the cross-border transportation route of forest chips from the Lendery terminal in Russia to Lieksa in Finland. Data on the average driving distance, average load of chip trucks, average timing of a work shift including idle time and breaks longer than 15 minutes were collected from 6 delivery runs in winter conditions. However, the situation in summer, regarding road conditions and other factors might be different, affecting the average driving speed, duration of the main operation and distribution of time within a work shift. In order to get a full picture of transportation productivity on the route throughout the whole year, it would be worth collecting and analysing summer data also. Demand for forest chips is relatively low in the summer and, according to data from the company, there were no deliveries at all in summer 2010 on the Lendery to Lieksa route. The reduced chip supply during the summer months, when the field work was done, made it difficult to obtain data on the chip trucks.

The chip truck drivers were interviewed to identify the main factors affecting the transportation productivity of their trucks. However, as well as the obstacles listed in the questionnaire, respondents referred to other factors which were not initially addressed. Unfortunately these factors were fully identified only later during the discussion with transportation company representatives. As it was possible to interview only a limited number of people from these companies, the overall picture regarding the main factors affecting productivity and their ranking might change radically if more respondents were to provide answers.

The representatives of the transport companies mentioned that the potential to increase productivity is limited by the current working schedule of the Inari border crossing point. According to Decree № 142 from 23.02.1994 with amendments from November 2007 (Article 5, paragraph 3), the supervision and regulatory work of temporary border crossing points should be organized by bilateral agreement, in which local authorities and interested parties should be involved. Currently there is demand from transport companies to improve the existing situation. However, at the moment it appears that the opinions and suggestions of interested parties, particularly transportation companies, are not adequately taken into account, and communication between the transportation companies and the local authorities is poor.

Therefore certain measures should be taken to improve the operation of the border crossing points. An open discussion process involving all interested parties would therefore help to solve the existing problem and find suitable solutions and compromises.

In addition to the infrastructure, there are other possibilities to improve the efficiency of the cross-border transportation of wood chips. Whereas the minimum reported unloading time was 30 minutes, unloading a modern chip truck with a 100 m^3 load is technically possible in from 30 seconds to 3 minutes, depending on the unloading system used (LIPE 2011). Proper organisation of the unloading process, e.g., decreasing the queueing time, could reduce the unloading time.

Due to the use of different reference years, the costs of forest chips transportation in Russia, in Finland and across the border were recalculated to a price level for 2010. For Finnish transportation costs, the initial data refer to 2003. It was not possible to find information on up-to-date transportation costs in Finland, because such information is a commercial secret. The calculated annual growth of Finnish transportation costs in this study is based on data from Tilastokeskus (2010) and the increase of transportation fuel index is 3% on average.

However, in the specific case of the transport by truck of wood logs, which could be considered logistically similar to the transport of forest chips, annual transportation costs for the same period grew on average by 5.7% (Statistical Year Book of Forestry 2009). This means essentially that the actual difference in transportation costs may be nearly double the figure used. This creates uncertainties in the cost recalculation and the probable solution for such a deviation might be to present cost changes in the form of different scenarios. However, another issue to address is how far statistics on wood log transportation costs are in fact applicable to forest chip transportation.

According to Hakkila (2004), the cost of road transportation of forest chips accounts for 35% of the total supply cost when using logging residues as a raw material, and 23% for whole-tree chipping. On average it therefore corresponds to 29% of the total supply cost. At the same time in Russian conditions the cost of road transportation of forest chips is 19% of the total supply cost for final felling and 17% on average for the first and second thinnings (Ilavský et al. 2007). It was found that the cost of cross-border transportation is 26% of the total supply cost.

Thus, the proportion of transportation costs in the total supply cost is highest for cross-border transportation at 26%, compared with 23% and 19% in Finland and Russia respectively. These costs were compared at a reference distance of 80 km.

The comparison in Figure 12 showed that the transportation costs of forest chips at 80 km reference distance is the highest within Finland (4.7 €/loose m^3), which is 26% higher than within Russia (3.5 €/loose m^3) and 28% higher than on cross-border transportation (3.4 €/loose m^3). But the cross-border transportation of forest chips is most expensive on the considered routes when the distance is less than 20 km. When the distance is greater than 20 km, their transportation from Russia to Finland is less expensive than within Finland. At 70 km distance the costs of cross-border transportation are equal to transportation costs in Russia and they are even lower at longer distances. This change in the cost-effectiveness of cross-border transportation can be explained by two factors. With short cross-border routes, the increased proportion of delays related to crossing the border affect the productivity of the trucks and consequently the transportation costs. The proportion of fuel costs in the total supply cost increase with distance, but the fuel VAT refund grows also, because transport companies engaged in cross-border transportation do not pay VAT. Also, it should be noted that drivers' salaries are lower in Russia.

In addition, transport companies involved in cross-border transportation of forest chips on this route buy fuel only in Finland to avoid risks related to the poor quality of Russian diesel. Thus, the cost disparity may be explained by the difference between fuel prices, labour costs and the maximum allowable weight of trucks. The road infrastructure and its condition should also be taken into account, as poor road conditions in Russia may cause additional costs.

The transportation distance for forest chips in the Leningrad region of up to 50 km (Goltsev et al. 2010) could be valid also for internal transportation of forest chips within Karelia; nevertheless cross-border transportation is in reality longer and the route studied here is 82 km. This may be explained by higher demand and consequently the better price paid by Finnish consumers.

The statistics provided by the transport company on the forest chip flow from the Lendery terminal to the Lieksa power plant showed that the supply chain is vulnerable due to its dependence on the availability of raw material at the terminal or breakdown of the chipper. Russian forest chips supply 10% of the total energy demand at Lieksa power plant. Storage space at the plant allows for the accumulation of raw material for a certain period of time and creates a buffer in supply in order to decrease the risks from any temporary cuts in deliveries.

As an alternative to transportation of forest chips from Russian Karelia to Finland, deliveries of non-merchantable raw material (e.g., firewood logs) could be made and chipping could be organized on the Finnish side. However, according to the interviews, such a scheme has already been tested by the companies and presented many challenges. In particular, one of the main problems was the Russian customs classification of firewood as industrial roundwood. This underlies the difference between customs fees, which for firewood are 4 €/m^3 with bark, and for industrial roundwood, e.g., pulpwood, are 15 €/m^3 with bark (Federal Customs Service of Russia 2009). Thus, this option is considered risky by the company, because it may be required to pay industrial wood customs fees when transporting firewood.

For cross-border transportation of forest chips, the customs fees are the lowest at only 5% of their value, taking into account that Russian forest chips are relatively cheap compared to those of Finnish origin. The price for Russian chips was at the lowest in 2009 when a lot of unclaimed pulpwood was chipped into forest chips on the Russian side and transported to Finland.

According to the interviews, the share of pulpwood in 2009 was up to 70% of the total amount of raw material used for chipping by the chip producers. Besides that, delays in payments by Russian pulp and paper companies encouraged forest chip producers to work with Finnish partners buying forest chips for energy purposes. The situation has changed as the general business environment in the Russian wood industry has improved in 2010, and deliveries to Finland are currently dependent upon the accumulation of a sufficient amount of non-

Table 7 Comparison of quality parameters of forest chips with different origin

Compared parameters	Logging residue chips (Impola 1998)	Roundwood chips (Impola 1998)	Forest chips obtained from Värtsila and Lahdenpohja terminals
Moisture content (%)	50-60	45-55	44-47
Net calorific value (dry matter) MJ/kg	18.5-20	18.5-20	18.4-18.7
Net calorific value of fuel as received MJ/kg	6-9	6-10	8.8-9.2
Particle size distribution (%): >20mm; 20-6.3; 6.3-3.15; <3.15	-	-	-
Ash content (%)	1-3	0.5-2	0.5-1.2

merchantable wood for chipping, which consequently again means risks for the continuity of forest chip transportation to Finland. In Finland logging residues are the main source of raw material for production of forest chips (Hakkila 2004), but in Russia forest chips are mainly produced from roundwood (Gerasimov and Karjalainen 2009b). In order to answer the question "can Russian forest chips substitute in terms of quality Finnish forest chips", the main quality parameters of forest chips obtained from Värtsilä and Lahdenpohja were compared with those produced in Finland from logging residues and roundwood (Impola 1998), as shown in Table 7.

Based on the data shown in Table 7 and Appendix A it can be concluded that the calorific value (dry matter and as received) and ash content for both Lahdenpohja and Värtsilä forest chips are within the intervals given by Impola (1998). The net calorific value of fuel as received from both terminals corresponds to the intervals provided by Impola (1998), which is the only reference to be formally compared, because the CEN/TS 14918 standard is currently under revision and minimum calorific value is yet to be stated. The moisture content fulfils the intervals for both categories of forest chip by Impola (1998); forest chips from Värtsilä correspond to the M45 class and forest chips from Lahdenpohja to the M50 class (CEN/TS 14961). The moisture content of samples from both locations is less than 50 % and meets the requirement of the K2 norm (Alakangas 2005), ranging from 41 to 50%. Particle size distribution of the samples from both locations conforms with the P2 norm (<45 mm) for 95% of all the particles (Alakangas et al. 2006). However, the biofuel standard has been updated lately (Alakangas 2010), and according to CEN/TS 14961 the particle size distribution of the obtained samples corresponds to P45B size class. According to the standard CEN/TS 14775, ash content for Lahdenpohja corresponds to class A0.5 and for Värtsilä to A1.5, and the ash content is within the intervals provided by Impola (1998). According to laboratory results, the samples of forest chips obtained from the terminals fulfil the quality requirement and European standard (CEN/TS 14961) for wood chips and log fuel. Thereby, Russian forest chips are of the same quality as Finnish one and they can substitute Finnish forest chips.

From the results of the laboratory tests, it was not easy to allocate the samples to the size classes due to discrepancy in the size of the sieving equipment used to determine particle sizes. The sieving equipment used for measurements of the particle size had only three sets of sieving screen apertures: 3.15, 6.3 and 20 mm, so it was possible to separate the chip particles into four size categories: <3.15, 3.15-6.3, 6.3-20 and >20 mm. However, size categories in the CEN/TS 14961 standard are distributed as follows: P16, P45, P63. For each category the particle size is as follows: 3.15≤P≤16 mm, 3.15≤P≤45 mm and 3.15≤P≤63 mm with the dominance of the main fraction for each class >80% by weight (CEN/TS 15149-1). Laboratory analysis showed that 6% of particle sizes for Lahdenpohja and 9% for Värtsilä are within the size category ">20 mm", although the distribution within interval of 16-20 mm is unknown. On the other hand, the suggested normal variation distribution of size makes it possible to indicate that the size class for both samples is P16, although it could be also stated with confidence that the samples fulfil the P45 size standard. Besides that, the developing standardisation of biofuels made it difficult to allocate quality characteristics of obtained samples to certain classes, e.g., to the CEN/TS 14961 standard, which consists of 6 parts, some of them currently under official voting for approval (Alakangas 2010). Therefore this standard can be considered only as one of the possible options, until it is officially approved. According to Ranta (2005), the transportation distance, particularly of forest chips, will be the main challenge for efficiency of fuel supply and it is likely to increase in future. As a result, the problems related to transportation cost-efficiency with low-energy intensity will be emphasised. Forest chips are mainly produced from by-products of logging and sawmilling and their procurement operation is quite often integrated with other forestry operations, such as procurement of roundwood or industrial wood waste (Asikainen 2001). The longest economically acceptable transportation distance for wood fuels in Finland has been less than 100 km, which is also valid for forest chips (Ranta 2005). At the same time, in Russia, 50 km is considered as the maximum distance for transportation of forest chips in terms of cost competitiveness with other fuels. The cross-border route studied is 82 km in total and is within the maximum limit of transportation in Finland. Results of the study showed that the transportation costs for forest chips delivered from the Lendery terminal in Russia to the Lieksa power plant in Finland are lower than the cost of forest chips transportation within Finland. In the study it was found that the costs of transportation of forest chips from Russia to Finland are affected by several factors, e.g., the productivity of chip trucks coming from Russia is lower than chip trucks working in Finland, at 90-110 and 120-140 loose m3 respectively. However, the cost-competitiveness of Russian forest chips in the case considered is supported by lower harvesting costs, lower average salaries in the forest sector and cheaper fuel for forest machinery. Besides that, the relatively weak position of low-quality woody biomass on the Russian domestic market creates a positive precondition for its export to Finland. The limited number of chip producers in Russian Karelia makes the supply chain vulnerable, however, and there are consequent risks related to that. For example, the volume of forest chip flow on the route considered decreased almost ninefold in the 2 months when the chipper at one of the terminals got broken. At the same time, the development of forest chip production in Russia to a large extent depends on external demand because the local demand for forest chips is quite low and conventional firewood is the predominant type of wood fuel used locally (Raitila et al. 2009). Besides, bigger settlements and municipalities in the Republic of Karelia are quite often connected to central natural gas distribution networks or use other fossil fuels for energy generation. The current weak support for the development of bioenergy in Russia's energy policy hinders the utilization of forest chips in spite of political initiatives related to renewable energy development in the Russian Federation (Energeticheskaya strategiya... 2003). However, in the Republic of Karelia the current policy related to promotion of bioenergy from local sources is promising, but still in the initial stage of development (Regional'naya celevaya programma... 2007, Regional'naya strategiya razvitiya... 2010). By contrast, the situation in Finland is different and there is a common understanding of the importance of bioenergy in the national energy sector and effective support measures have been implemented (Puun energiakäyttö 2010), which also facilitate further development of cross-border trade of forest chips. The results obtained revealed the difference in average speed within the transportation route studied. The difference is almost twofold, 34 km/h in Russia and 66 km/h in Finland. The transportation distance on the Finnish side is 57 km and on the Russian side 25 km, but in terms of the time structure of a run, the proportion of driving on the Finnish side is 27% and on the Russian side 22%, which is again connected to the average driving speed on both sides. The difference is due to poorer road conditions on the Russian part of the route. The importance of this factor was proven during the identification of the main driving factors influencing the efficiency of forest chip transportation. The respondents indicated that bad road conditions on the Russian side and idle time at the border were among the most important factors influencing cross-border transportation. In addition, other factors affecting transportation productivity, which were not reflected in the questionnaire, were identified during communication with the transport companies. For example, the current working schedules of the Inari border crossing point and the inaccessibility of roads in Russia for at least two months of the year were mentioned by the companies' representatives as factors with a strong impact on transportation productivity. All these factors together decrease the overall productivity of cross-border transportation of forest chips. In the case considered, only 8 of 10 planned round trips by the chip truck were made during one week. The quality characteristics of the obtained forest chip samples from both terminals in Russian Karelia are similar to those of Finnish forest chips given by Impola et al. (1998). The characteristics of Russian forest chips fulfil the requirements of the quality standards for forest chips "CEN/TS 14961 – fuel specification and classes". Most forest chips produced in Russia are made of roundwood, while in Finland they are made mainly from logging residues or small diameter trees, so procurement costs may differ between the two sources. Taking into account the quality characteristics and costs, it may be concluded that Russian forest chips are cost-competitive and can compete to some extent with Finnish chips. The competitiveness of Russian forest chips is limited but can be improved. The analysis of domestic and cross-border transportation costs showed that the proportions of the transportation costs in the total supply cost of forest chips are different in all the cases considered. Of the total supply costs of forest chips, the highest proportion of transportation costs were on the cross-border route (26%), while transportation costs were a smaller proportion of the total supply costs in Russia and Finland (19% and 23% respectively). The high proportion of transportation costs in the case of the cross-border route is significant. According to the results of this study, the reasons are poorer road conditions on the Russian side, idle time at the border crossing point and differences between transport legislation in Russia and Finland. The impact of the state of Russian roads on productivity of cross-border transportation of forest chips was revealed when analysing the chip truck runs and further comfirmed in the interviews with the chip truck drivers. Properly addressed measures can increase overall transportation productivity and reduce transportation costs. Consequently, transportation distances can be expanded to provide large procurement areas for users of Russian forest chips. It is necessary to invest more in the development of the road infrastructure in Russia and improve the border crossing customs formalities and legislation.

ACKNOWLEDGMENTS

This study was done within the project "Wood Harvesting and Logistics in Russia – Focus on Research and Business Opportunities". The project studies the current state and the future development of wood procurement in Northwest Russia to strengthen the business positions of the project's stakeholders on the emerging Russian market. The project is financed by Tekes, the Finnish Funding Agency for Technology and Innovation, and a consortium of Finnish companies.

Appendix A

Table 8 Comparative analysis of quality of forest chips with different standards

Compared parameters	Logging residue chips (Impola 1998)	Roundwood chip (Impola 1998)	Lahdenpohja forest chips	Värtsila forest chips	Standard
					CEN/TS 14961
Moisture content (%)	50-60	45-55	47	44	M50 (Lahdenponja) M45 (Värtsilä)
Net calorific value (dry matter) MJ/kg	18.5-20	18.5-20	18.7	18.4	-
Net calorific value of fuel as received MJ/kg	6-9	6-10	8.8	9.2	-
Particle size distribution (%): >20mm; 20-6.3; 6.3-3.15; <3.15	-	-	6, 66, 20, 8	7, 70, 11, 10	P45B (Lahdenpohja, Värtsilä)
Ash content (%)	1-3	0.5-2	0.5	1.2	A0.5 (Lahdenpohja) A1.5 (Värtsilä)

References

Alakangas, E. 2005. Properties of wood fuels used in Finland, Technical Research Centre of Finland, VTT Processes, Project report PRO2/P2030/05 (Project C5SU00800), Jyväskylä 2005, 90 p. + app. 10 p.

Alakangas, E. 2010. European standard (EN 14961) for wood chips and hog fuel. VTT Technical Research Centre of Finland. *Forest Bioenergy 2010 – Book of Proceedings*, p. 329-341.

Alakangas, E., Valtanen, J. and Levlin, J.E., 2006. CEN technical specification for solid biofuels - Fuel specification and classes, Biomass and Bioenergy, Volume 30, Issue 11, *Standarisation of Solid Biofuels in Europe, Standarisation of Solid Biofuels in Europe*, ISSN 0961-9534. Available at: http://www.sciencedirect.com/science/article/B6V22-4KRY3MK-1/2/2 0549 7a18d 82752e06556f99fb28fef2

Asetus ajoneuvojen käytöstä tiellä annetun asetuksen muuttamisesta. §23 Auton ja perävaunun yhdistelmän koko- naismassa. 1997 Available at: http://www .finlex.fi/fi/laki/alkup/1997/19970670

Asikainen, A. 2001. Design of supply chains for forest fuels. In: Sjöström, K. (ed.). *Supply chain management for paper and timber industries*. Proceedings of the 2nd World Symposium on Logistics in Forest Sector, 12–15 August 2001, Växjö, Sweden. p. 179–190.

Bing maps. *Microsoft Corporation* 2010. Available at: http://www.bing.com

CEN/TS 14774-1, *Solid biofuels – Methods for the determination of moisture content – Oven dry method* - Part 1: Total moisture – Reference method.

CEN/TS 14774-2, *Solid biofuels – Methods for the determination of moisture content – Oven dry method* – Part 2: Total moisture – Simplified procedure

CEN/TS 14775, *Solid biofuels – Determination of ash content.*

CEN/TS 14778-1, *Solid Biofuels – Sampling* – Part 1: Methods for sampling.

CEN/TS 14918, *Solid Biofuels – Method for the determination of calorific value.*

CEN/TS 14961, *Fuel specifications and classes.*

CEN/TS 15149-1, *Solid biofuels - Methods for the determination of particle size distribution* - Part 1: Oscillating screen method using sieve apertures of 3,15 mm and above.

Decree № 142 from 23.02.1994 with amendments from November 2007 (Article 5, paragraph 3). *Russian-Finnish bilateral agreement regarding border crossing points* (In Russian.) Available at: http://lawrussia. ru/texts/legal_383/doc383a720x887.htm

Devyatkin, V. (ed.-in-chief). 1999. *Sbornik udel'nich pokazateley obrazovaniya otxodov proizvodstva i potrebleniya* [Collection of waste generation indexes in production and consumption]. Moscow. State Committee of Russian Federation for Environment. 65 p. (In Russian.)

Energeticheskaya strategiya Rossii na period do 2030 goda [Energy strategy of Russian Federation utill 2030]. Utverzdena rasporyazeniem Pravitelstva Rossiyskoy Federacii ot 13 noyabrya 2009 goda № 1715-r. (In Russian.)

Eurostat 2009. *Demographic Outlook – National reports on the demographic developments in 2007*. Luxembourg: Office for Official Publications of the European Communities. 65 p.

Federal Agency for Development of the Borders of the Russian Federation. 2009. Rabochaya gruppa Rosgranici izuchaet opit Finlandii po organizacii I funkcionirovanuyu punktov propuska [The working group of Rosgranitsa is learning the Finnish experience of management of border crossing points]. Avaialble at: http://www.rosgranitsa.ru/node/1301

Federal Customs Service of Russia 2009. Export statistics. Available at: www.customs.ru

Federal Service of State Statistics. 2010. Indeksi tarifov na gruzoviye perevozki osnovnimi vidami transporta. Procent. Leningradskaya oblast'. Avtomobil'niy. Dekabr'[Federal Department of Statistics. Indexes of main transportation tariffs. Percentage. Leningrad region. Motor-car. December.] (In Russian.) Available at: http://www.gks.ru/ dbscripts/cbsd/dbinet .cgi?pl =1934004

Gerasimov, Yuri and Karjalainen, Timo. 2009a. Estimation of supply and delivery cost of energy wood from Northwest Russia. Metlan työraportteja / Working Papers of the Finnish Forest Research Institute 123. 21 p. ISBN 978-951-40-2166-4 (PDF). Available at: http://www.metla.fi/julkaisut/ workingpapers /2009/mwp123.htm

Gerasimov, Yuri and Karjalainen, Timo. 2009b. Assessment of Energy Wood Resources in Northwest Russia. Metlan työraportteja / Working Papers of the Finnish Forest Research Institute 108. 52 s. ISBN 978-951-40-2147-3 (PDF). Saatavissa: http://www.metla.fi/julkaisut/workingpapers/2009/ mwp108.htm

Gerasimov, Yuri, Karvinen, Sari and Leinonen, Timo. 2009a. Atlas of the forest sector in Northwest Russia 2009. Metlan työraportteja / Working Papers of the Finnish Forest Research Institute 131. 43 p. ISBN 978-951-40-2183-1 (PDF), ISBN 978-951-40-2184-8 (paperback). Available at: http://www.metla .fi/julkaisut/workingpapers/2009/mwp131.htm

Gerasimov, Yuri, Sibiryakov, K, Moshkov, S, Välkky, Elina, Karvinen, Sari. 2009b. Raschet ekspluatacionnych zatrat lesosechnych mashin. Finnish Forest Research Institute. ISBN 978-951-40-2174-9. Available at: http://www .idanmetsatieto.info/rus/cfmldocs/index.cfm?ID=683andtiedote=viewandtied ote_ID=1799.

Goltsev, Vadim, Ján Ilavský, Yuri Gerasimov and Timo, Karjalainen. 2010. Potential biofuel development in Tikhvin and Boksitogorsk districts of the Leningrad region – The analysis of energy wood supply systems and costs. *Forest Policy and Economics* 12, no. 4 (2010): 308-316.

Goltsev, Vadim, Maxim Trishkin, and Timo Tolonen. "Efficiency of forest chip transportation from Russian Karelia to Finland". Joensuu, Finland, 2011. Joensuu, Finland. http://www.metla.fi/julkaisut/workingpapers/ 2011/mwp 189.htm.

Google maps. Google, Geocentre Consulting, Tele Atlas 2010. Available at: http://maps.google.com/

Grigoryev, M. 2007. *Povishenie roli mestnych toplivo-energeticheskich resursov v obespechenii energeticheskoy bezopasnosti Severo-Zapada Rossii* [The increasing role of domestic fuel-ener- gy resources for energy security in Northwest Russia]. Gazovy Business, July-August 2007: 28-34. (In Russian.)

Hakkila, P. 2004. Developing technology for large-scale production of forest chips. *Wood Energy Technology Programme 1999—2003*, *Technology Programme Report* 6/2004. Final report. Helsinki 2004. 99 p. Available at: www.tekes.fi/english/programm/woodenergy

Hakkila, P., 2006. Factors driving the development of forest energy in Finland. *Biomass and Bioenergy*, 30(4), 281-288. Available at: http://linkinghub. elsevier.com/ retrieve/pii/S0961953405001546

Heinimo, J., 2008. Methodological aspects on international biofuels trade: International streams and trade of solid and liquid biofuels in Finland. Biomass and Bioenergy, 32(8), 702-716. Available at: http://linkinghub .elsevier.com/retrieve/pii/S0961953408000111

Ilavský, Ján, Goltsev, Vadim, Karjalainen, Timo, Gerasimov, Yuri and Tahvanainen, Timo. 2007. *Energy Wood Potential, Supply Systems and Costs in Tihvin and Boksitogorsk Districts of the Leningrad Region*. Metlan työraportteja / Working Papers of the Finnish Forest Research Institute 64. 37 p. ISBN 978-951-40-2074-2 (PDF). Available at: http://www.metla.fi /julkaisut/workingpapers/2007/mwp064.htm.

Impola, R. 1998. *Puupolttoaineiden laatuohje*. (Quality instructions for wood fuels). Jyväskylä, FINBIO, Publication 5. 33 p.

Instrukciya po perevozke krupnogabaritnich i tyazelovesnich gruzov avtomobil'nim transportom po dorogam Rossijskoy Federacii. 1996 [Instruction on transportation of large-sized and heavy loads by trucks in Russian Federation] (In Russian).

Kareliyastat 2009. Республика Карелия в цифрах. Население. Численность постоянного населения. [Population in Republic of Karelia]. [Online Document]. Available at: http://krl.gks.ru/digital/region1/DocLib/ dem1.htm

Kareliyastat 2010. *Eksport schepy i toplivnoy drevesini iz Respubliki Karelia*. [Export of wood chips and fuel wood from the Republic of Karelia]. 2 p.

Kareliyastat. 2008. Lesopromishlenniy kompleks regionov Severo-Zapadnogo okruga Rossii [Forest sector of the federal regions of Northwest Russia]. Petrozavodsk: Karelian Branch of Federal State Statistics Service (Kareliastat). 201 p. (In Russian).

Kofman, P.D. 2007. Ordering and receiving wood chip fuel, COFORD Connects Note Processing/Products nr 8, Dublin, 2007.

Lindblad, Jari, Äijälä, Olli, Koistinen, Arto, 2008. Energiapuun mittaus. Available at: http://www.metla.fi/metinfo/tietopaketit/mittaus/aineistoja/ energiapuun-mittausopas-2008.pdf

LIPE 2011. Technical specification of the LIPE chip trailers. Available at: http://www.anttiranta.com/pdf/esite/ESITE_LIPE_ENG.pdf

MED, 2011. Average prices on wood and wood fuels. Petrozavodsk, Russia: The Ministery of Economic Development of the Republic of Karelia, 2011. Petrozavodsk, Russia. Available at: http://www.gov.karelia.ru/ gov/Power /Ministry/Development/Prices/price_wood_1012e.html.

Ministry of the forest complex of the Republic of Karelia, 2010. Official request on data obtaining to the Ministry.

Ministry of Trade and Industry, 2000. *Action plan for renewable energy sources.* Publications 1/2000.

OECD/IEA, 2003. *Renewables in Russia - From Opportunity to Reality.* International Energy Agency, 2003. ISBN 92-64-105441.

Palvelujen ulkomaankaupan arvonlisäverotus 1.1.2010 alkaen. 2010. Verohallinto. Arvonlisäverotus/Kansainvälinen kauppa. Available at: http:/ /www.vero.fi/?article=2483anddomain=VERO_MAINandpath=5,40andlangu age=FIN.

Puun energiakäyttö 2009. Toimittaja: Esa Ylitalo. Metsähakkeen käyttö nousi yli 6 miljoonan kuutiometrin. Metsäntutkimuslaitos, Metsätilastollinen tieto palvelu. *Metsätilastotiedote* 16/2010, 28.4.2010

PÖYRY. "Polttoaineiden hintataso." *Bioenergia*, no. 4 (2010): 48.

Raitila, J., Virkkunen, M., Flyktman, M., Leinonen, A., Gerasimov, Y. and Karjalainen, T. 2009. *The pre- feasibility study of biomass plant in Kostomuksha*. Research Report VTT-R-08372-08. 54 p.

Rakitova, Olga, Anton Ovsyanko, Richard Sikkema, and Martin Junginger. Wood Pellets Production and Trade in Russia , Belarus and Ukraine, 2009. http://www.forcebioenergy.dk/pelletsatlas_docs/showdoc.asp?id=090520131 636andtype=docandpdf=true.

Ranta, T. 2002. *Logging residues from regeneration fellings for biofuel production - a GIS-based availability and supply.* Thesis for the degree of Doctor of Science. Lappeenranta University of Technology.

Ranta, T. 2005. Logging residues from regeneration fellings for biofuel production–a GIS-based availability analysis in Finland. *Biomass and Bioenergy*, Volume 28, Issue 2, February 2005, Pages 171-182.

Regional'naya celevaya programma "Aktivnoe vovlechenie v toplivno-energeticheskiy kompleks Respubliki Karelia mestnich toplivno-energe ticheskich resursov na 2007-2010 godi [Regional programme aimed on intensive utilization of local fuels in Republic of Karelia in 2007-2010]. The Republic of Karelia State Government Bodies' Official Web Portal, 2010. Available at: http://www.gov.karelia.ru/gov/News/2010/ 09/0920_02.html.

Regional'naya strategiya razvitiya toplivnoy otrasli Respubliki Karelia na osnove mestnich energeticheskich resursov na 2011-2020 godi [Regional strategy on development of energy sector in Republic of Karelia in 2011-2020]. The Republic of Karelia State Government Bodies' Official Web Portal of the Government of the Republic of Karelia, 2010. Available at: http://www.gov.karelia.ru/gov/News/2010/09/0920_02.html

Renewable energy policy review. Finland 2009. In the framework of the EU co – funded project: RES 2020: Monitoring and Evaluation of the RES Directives implementation in EU27 and policy recommendations to 2020.

Roslesinforg. 2008. *Svedeniya gosudarstvennogo statisticheskogo nablyudeniya po lesopolzovaniyu za 2007 god* [Information about the state statistical monitoring on forest use in 2007]. (In Russian.)

Syunev, Vladimir, Sokolov, Anton, Konovalov, Alexandr, Katarov, Vasily, Seliverstov, Alexandr, Gerasimov, Yuri, Karvinen, Sari and Välkky, Elina. 2009. Comparison of wood harvesting methods in the Republic of Karelia. Metlan työraportteja / Working Papers of the Finnish Forest Research Institute 120. 117 p. ISBN 978-951-40-2162-6 (PDF). Available at: http://www.metla.fi/julkaisut/workingpapers/2009/mwp120. htm

The republic of Karelia in brief. The Republic of Karelia State Government Bodies' Official Web Portal. Available at: http://www.gov.karelia.ru/ gov/Different /karelia3_e.html.

The Republic of Karelia State Government Bodies' Official Web Portal, 2010. Lesnoy otrasli Karelii udalos' spravitsya s trudnostyami [The forest sector of Karelia has overcome the obstacles] Available at: http://www.gov. karelia.ru/gov/News/2010/02/0218_03.html

Tilastokeskus, 2010. Kuorma-autoliikenteen kustannukset nousivat vuodessa 2,7 procenttia. Kuorma-autoliikenteen kustannusindeksi. Available at: http:/ /www.tilastokeskus.fi/til/kalki/2009/12/kalki_2009_12_2010-01-18_tie_001_fi.html

INDEX

C

F

G

H

I

J

K

L

M

N

O

P

Q

R

S

T

U

V

W

X

Y

Z